J. L. Cloudsley-Thompson

Ecology and Behaviour of Mesozoic Rep

John L. Cloudsley-Thompson

Ecology and Behaviour of Mesozoic Reptiles

With 129 Figures and 4 Tables

Springer

Professor Dr. John L. Cloudsley-Thompson
10 Battishill Street
London, N1 1TE
UK

ISBN 3-540-22421-1 Springer-Verlag Berlin Heidelberg New York

Library of Congress Control Number: 2004112900

Springer is a part of Springer Science + Business Media
springeronline.com

Springer-Verlag Berlin Heidelberg 2005
Printed in Germany

Editor: Dr. Dieter Czeschlik, Heidelberg
Desk editor: Dr. Andrea Schlitzberger, Heidelberg
Cover design: Design & Production, Heidelberg
Cover illustration: *Rhamphorhynchus* (see Fig. 58)
Typesetting: perform electronic publishing GmbH, Heidelberg
31/3150WI – 5 4 3 2 1 0 – Printed on acid-free paper

Preface

Popular interest in dinosaurs and other Mesozoic reptiles began during the third decade of the 19th century, and has increased continually since that time. In 1841, the name Dinosauria ('terrible lizards') was proposed by Richard Owen at the annual meeting of the British Association for the Advancement of Science, held in Plymouth; it first appeared in print the following year. During subsequent decades, both scientific and popular interest were extended to pterosaurs, mosasaurs, ichthyosaurs and plesiosaurs. Mesozoic reptiles, especially dinosaurs, likewise earned their place in science fiction, films and comic strips. It is no coincidence that both my wife and I were thrilled especially by Sir Arthur Conan Doyle's *The Lost World* (1912) whilst at school. Anne even read it by the flickering light of a coal fire whilst in the sanatorium quarantined for chickenpox! Edgar Rice Burroughs, creator of Tarzan, and numerous other authors wrote stories about dinosaurs during the first part of the 20th Century, whilst they appeared in many early films, notably *The Lost World* (1925) and *King Kong* (1933).

My own interest in Mesozoic reptiles was awakened at the age of 6 or 7 when my parents first took me to the Natural History Museum, London. I was astonished by their wondrous size and the sobering thought that they had all disappeared from the face of the earth some 65 million years ago. When we got home I asked my father, What "What if there was a dinosaur buried underneath our house?" I never forgot his reply, "Then the house would have to be pulled down to get it out." As I grew older, the mystery of dinosaur extinction intrigued me more and more. On my return to Cambridge after World War II, despite a very busy timetable and knowing that I would not attempt to answer any questions on vertebrate palaeontology in the Natural Science Tripos Part 2 exam, I nevertheless attended F.R. Parrington's lectures, as did Alan Charig. Indeed, the Cretaceous extinction did not impinge upon my own fields of research for 30 years. By that time, a number of clues had been provided that helped to explain the mystery . They came not only from palaeontological and geological studies, but also from research on the physiology and ecology of extant reptiles.

In the present volume, I have attempted to provide a not unbiased synthesis of current views regarding the ecology, physiology and behaviour of Mesozoic reptiles, and to outline the various hypotheses that have been proposed to explain their extinction.

London, Autumn 2004 John Cloudsley-Thompson

Acknowledgements

My warmest thanks to Prof Dr Michael Benton, author of over 40 books on evolution, fossils and dinosaurs, for invaluable advice and for his kindness in allowing me to plunder his much acclaimed *Vertebrate Palaeontology.* I should also like to express my appreciation for the help and advice of Laura Tatham who typed the manuscript and of Michael Palmer, archivist to The Zoological Society of London, who traced some elusive references for me. Once again, my thanks to Dr Dieter Czeschlik and Dr Andrea Schlitzberger, Biology Editorial Springer-Verlag, for their friendly cooperation. As always, I am deeply indebted to my dearest wife Anne for her wonderful companionship and advice. Alison Bailey, on behalf of the Editorial Board, kindly gave me permission to quote from an article of mine that first appeared in *Biologist* (Institute of Biology) in 2001. I am also grateful to various artists, authors and publishers for granting permission to redraw and adapt copyright material. Sources have been credited both in the text and in the legends to the figures; full details are cited in the bibliography. Every reasonable effort has been made to trace and obtain permission but, if any copyright owners have inadvertently been overlooked, I will be pleased to make the normal arrangements and give full acknowledgements at the first opportunity.

Contents

1 Classification

1.1 Introduction

The Mesozoic Era is popularly known as 'The Age of Reptiles'. It comprises three periods: Triassic, Jurassic, and Cretaceous – but reptiles were already present during the preceding Carboniferous and Permian periods. These were the last periods of the Palaeozoic Era (Table 1). (Palaeozoic means 'ancient life', Mesozoic 'middle life' and Cenozoic, which followed it and continues to the present, 'recent life'.)

Life probably began in the sea, and terrestrial plants and animals did not appear until the Ordovician period: this followed the Cambrian and preceded the Silurian. Terrestrial vegetation during the Ordovician consisted mainly of simple, non-vascular plants (which probably included mosses, liverworts, lichens, and fungi), while the fossil soils contain burrows presumably made by worms and arthropods. Land plants diversified during the Silurian period, while the fauna consisted mainly of millipedes, centipedes, and tiny, fungus-eating microarthropods. Thus, even then, there was the beginning of a terrestrial food web consisting of primary producers, decomposers, secondary consumers, and predators. Nearly all the terrestrial animals of the Devonian period were arthropods. The phylum Arthropoda diversified considerably at that time; mites and springtails (Collembola) appeared, and the first amphibians crawled onto the land (Benton and Harper 1997).

Table 1. Periods of the Palaeozoic, and Mesozoic Eras

Period	Approximate date (millions of years) before the present	Approximate duration (millions of years)
Cretaceous	145–65	80
Jurassic	205–145	60
Triassic	250–205	45
Permian	270–250	20
Carboniferous	350–270	80
Devonian	400–350	50
Silurian	440–400	40
Ordovician	480–440	40
Cambrian	550–480	70

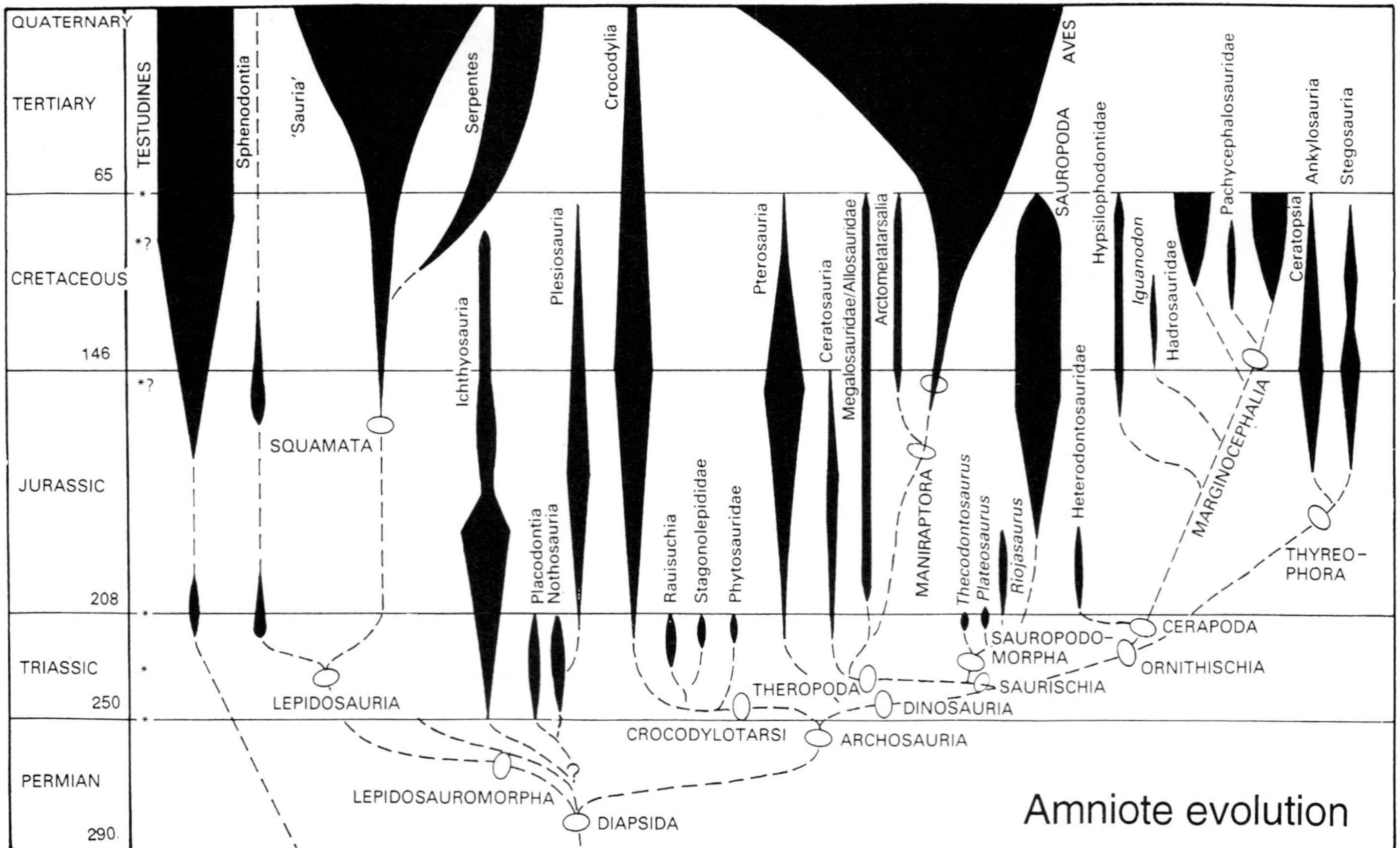

Fig. 1. Phylogenetic tree of the diapsid reptiles and turtles, showing postulated phylogenetic relationships of the main groups (*dashed lines*), known fossil records of each (vertical time dimension) and their relative abundance through time (horizontal axis). Mass extinctions and possible mass extinctions are noted with *asterisks* on the left. (Benton 2004)

1.2 Amphibians

Amphibians evolved from the lobe-finned fishes (Rhipidistia) that inhabited shallow freshwaters during the Devonian period. Several types of amphibian appeared, of which one group, the Anthracosauria, gave rise to the reptiles. There is such a close resemblance between the skulls of the Devonian rhipidistian fishes and those of the earliest amphibians that their relationship is not in doubt.

The Palaeozoic amphibians were of two main types, consisting of the orders Temnospondyli ('divided vertebrae') and Anthracosauria ('coal lizards'). The latter contained forms from which the Reptilia evolved. Modern Amphibia are generally considered to be a single monophyletic group and probably evolved from the Temnospondyli. (Monophyly is the condition in which a group or taxon of related organisms shares a common ancestry.)

Placed in the subclass Lissamphibia ('smooth amphibians'), existing or extant amphibians do not enjoy any close affinity with reptiles. Although they possessed lungs, the ancestors of modern Amphibia must have lost their scales and evolved cutaneous respiration through moist skins. This can be deduced from the fact that they could not have expanded their lungs by means of ribs. Instead, they forced air down their throats with the aid of special muscles and the bones which supported the floor of the mouth, as existing lissamphibians do. When the floor of the mouth of a frog or a salamander is lowered, air is sucked in (Cloudsley-Thompson 1999). The nostrils are then closed and the throat muscles force this air down into the lungs. The muscles are then relaxed, allowing the air to be released. It is highly unlikely that such a method of breathing would have been adequate for active life on land unless supplemented by accessory cutaneous respiration.

No fossils have yet been found of the amphibian group immediately ancestral to reptiles, or of the primitive reptilian stock prior to its initial diversification (Currie and Padian 1977; Farlow and Brett-Surman 1977). The oldest reptilian fossils known are of the ancestral mammal-like reptiles and of Anapsida (Fig. 1). The latter are usually regarded as being the more primitive on account of their small size and unspecialised skeletons. They gave rise to the diapsids (see below) which include the majority of reptilian groups, both fossil and extant.

1.3 Reptiles

The first reptiles – forms like *Hylonomus* (Sect. 2.4) appeared in the Middle Carboniferous (Benton 1996). These shortly led to the three main divisions of reptiles, the anapsids, diapsids and synapsids, characterised by the temporal openings or fenestrae in their skulls (Fig. 2). Indeed, knowledge of the interrelationships of reptiles depends mainly upon their fossil skeletons, of which skulls are by far the most useful and important. They are of four different types (Fig. 2). In the subclass Anapsida, the dermal bones of the temporal region of

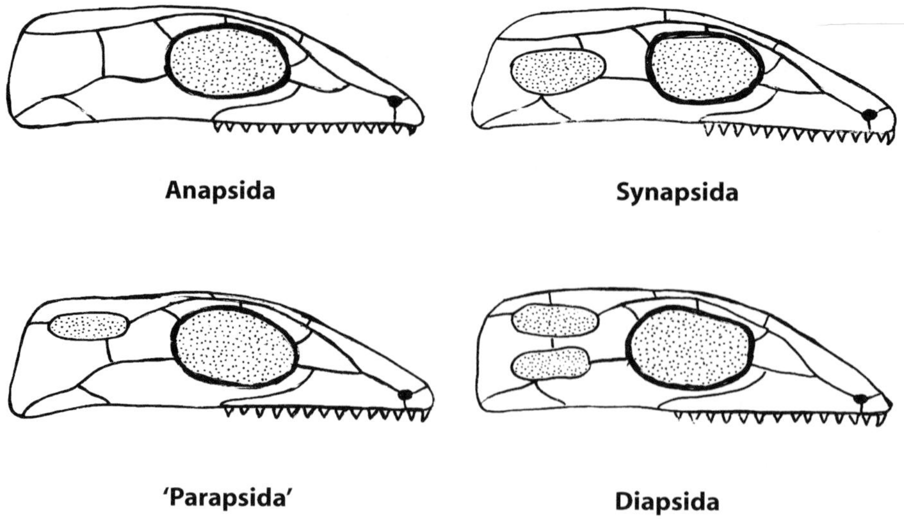

Fig. 2. Reptilian skulls showing arrangement of temporal openings. The various bones have not been named. (Cloudsley-Thompson 1999 based on Young 1981)

the skull behind the orbit present an unbroken surface and there are no apertures between them. This character separates the parareptiles ('beside reptiles') and the Testudines or Chelonia (turtles and tortoises) from all other reptiles. In more advanced reptiles, there are one or two fenestrae surrounded by bony arches in the temporal region. These holes allow the jaw muscles to bulge into them, thereby facilitating their actions during feeding. The Lepidosauria and Archosauria, probably the most successful subclasses of reptiles, have two fenestrae and are therefore known as Diapsida. The lower temporal arch is incomplete in the Squamata, while the upper arch, too, has been lost in some extant lizards and in snakes (Table 2). Only a single fossa and arch are present in the remaining subclasses. In parapsid reptiles this is situated high on the skull. This condition is found in the Ichthyosauria and Plesiosauria, both of which probably evolved from diapsid precursors. In the remaining subclass, Synapsida, there is also only one fenestra but, in the earlier forms at least, this was located lower in the skull (Fig. 2). The term synapsid means fused arch. It was applied to the group by early workers who believed, incorrectly, that the single arch had been derived from fusion of the two arches found in diapsid reptiles (Table 2).

When Table 2 is considered in relation to Figs. 1 and 2, the relationships between the various taxa or groups of reptiles should become clear. The only reptiles found in the world today are Testudines or Chelonia, Lepidosauria, and Crocodylia. The remainder are extinct.

One final point requires clarification. It will be seen that the dinosaurs ('terrible lizards') incorporated two orders, Saurischia and Ornithischia. The Sauri-

Table 2. Simplified classification of reptiles

Class Reptiles		
Subclass Anapsida ('without arches')		
	Order Parareptilia ('beside reptiles')	
	Family Mesosauridae ('middle lizards')	
	Order Testudines or Chelonia[a] (turtles and tortoises)	
Subclass Sauropterygia ('reptile fins')		
	Placodonts, nothosaurs, plesiosaurs etc.	
Subclass Ichthyopterygia ('fish fins')		
	Ichthyosaurs ('fish lizards')	
Subclass Diapsida ('double arches')		
Infraclass Lepidosauria ('scaly lizards')		
	Order Rhynchocephalia[a] ('beak headed')	
	Order Squamata[a] ('scaly ones')	
		Lizards, snakes, and amphisbaenians
Infraclass Archosauria ('ruling reptiles')		
	Order Thecodontia ('socketed teeth')	
	Order Crocodylia[a] (crocodiles)	
	Order Pterosauria ('wing lizards')	
		Pterodactyls, etc
Superorder Dinosauria ('terrible lizards')		
	Order Saurischia ('lizard pelvis')	
		Theropod and sauropod dinosaurs
	Order Ornithischia ('bird pelvis')	
		Ornithopod dinosaurs. Stegosaurs, ankylosaurs, and ceratopsians
Subclass Synapsida ('fused arches')		
	Order Pelycosauria ('basin lizards')	
	Order Therapsida ('mammal arches')	
		Mammal-like reptiles

[a] Groups containing extant representatives

schia comprised two suborders Theropoda and Sauropoda. The earlier therospods were small, bipedal dinosaurs but, by the end of the Cretaceous period, they had produced *Tyrannosaurus* and its relatives, the largest carnivores ever to be seen on land. The Sauropoda, on the other hand, were quadrupedal vegetarians culminating in the immense Jurassic *Apatosaurus, Diplodocus,* and *Seismosaurus.* The second order of dinosaurs, the Ornithischia, appeared in the Jurassic and reached a peak in the Cretaceous period, by which time the sauropods had already become less common. The first ornithischians were bipedal, like *Iguanodon*, but several lines reverted to a quadrupedal habit and some of the latter developed heavy armour and defensive spines.

The main difference between the two orders lies in the shape of the pelvis (Fig. 3). In the Saurischia, this was similar to that of other reptiles, but in the Ornithischia it was bird-like. In both orders, the hip socket was surrounded by

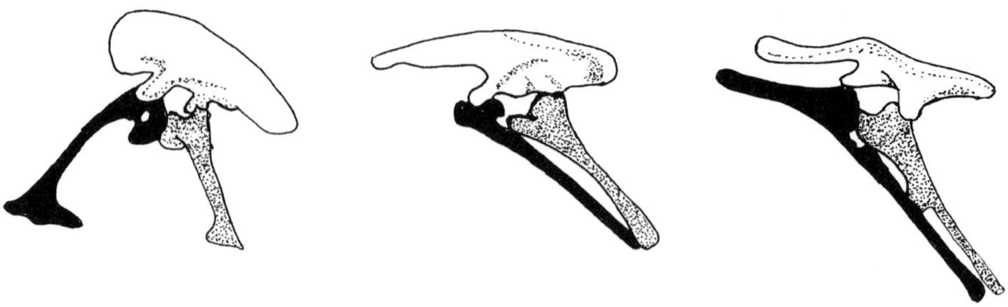

Fig. 3. Pelvic bones of dinosaurs. *Left* A saurischian; *centre* a primitive ornithischian; *right* a typical ornithischian. Ilium (*white*); ischium (*stippled*); pubis (*black*). (Cloudsley-Thompson 1999)

three bones: the ilium above, the ischium below and behind, the pubis below and in front. In the Saurischia the pubis pointed forward and downwards as a single prong; while, in the Ornithischia, it usually had two prongs, one projecting forward and upwards, the other backwards so that it lay just below the ischium. In certain primitive ornithischians the forward prong was absent.

The Triassic archosaurs gave rise to the pterosaurs and the dinosaurs: the saurischian dinosaurs gave rise to birds in the Jurassic period. Recent sequencing of whole mitochondrial genomes suggests that turtles may be related to a bird/crocodile clade or linear branch. (The term clade is derived from a Greek word meaning a twig or branch.) Birds are much more closely related to Mesozoic reptiles than mammals are, and the Testudines may be more like diapsids than has previously been realised (see Benton 1990a).

The most bird-like of the dinosaurs were the Cretaceous theropods, whereas birds were well established by the end of the Jurassic period. Birds flourished during the Mesozoic Era and colonised numerous ecological niches. The number of known fossilised bird taxa has increased greatly in recent years, shedding light on the large temporal and evolutionary gap between the Late Jurassic *Archaeopteryx lithographica* on the one hand, and superficially diver-like hesperornithiform and pelican-like ichthyornithiform birds on the other. These findings provide additional evidence for the hypothesis that birds evolved from cursorial, bipedal carnivorous dinosaurs during the Jurassic period.

Although doubts have been expressed regarding the hypothesis that birds evolved from dinosaurs, the discovery that the microscopic Haversian canals (which carry nutrients to bone cells) of coelurosaurian dinosaurs were organized in the same way as they are in birds today, suggests a strong link between the two. Furthermore, the bundles of collagen fibres which bind bone materials have the same irregular structure in both birds and coelurosaurians. These matters will be discussed further in Chapter 11 (Sect. 11.5).

Our knowledge of extinct animals depends almost entirely upon the study of fossils, as already mentioned. Usually only the hardest parts of animals, such as bones and shells, become fossilised. It should be remembered that countless generations of innumerable species have existed in the past without leaving any traces, and that the fossils we do have represent only a minute fraction of the billions of reptiles that populated the Mesozoic Era. Yet, it is with these that palaeontologists work to reconstruct individual skeletons and even entire extinct faunas. The object of the present book is to clothe these skeletons with flesh, and to show something of the ways in which their possessors lived.

Taxonomy is the scientific ordering and labelling of groups of similar organisms, while systematics is the study of the diversity of organisms within clades. Clades are groups of organisms that are monophyletic – that is, they have a common ancestry. Such groups are known as taxa and are assigned a Linnaean rank. Standard ranks are phylum, class, order, family, genus and species. Between these are intermediate groups and subgroups at different levels, e.g. superclasses, subfamilies and so on. In addition, divisions, cohorts and other ranks may be added. This system was the one most widely used until the 1970s and has the advantage of showing which taxa are most closely related, and it indicated their hierarchical levels. For convenience it has been generally adopted in the present book.

It must be realised, however, that an order such as Thecodontia (Table 2) includes archosaurs that are more advanced than earlier reptiles, but more primitive than dinosaurs and birds. At the same time, they are not characterised by any unique features and they lack the specialisations of the more advanced forms. Cladistics, or phylogenetic systematics, is therefore now used by most taxonomists in place of the earlier evolutionary systematics, because it shows the closeness of the ancestry between groups that arose when populations split – a matter of little concern here (Holtz and Brett-Surman 1997). For a detailed cladistic classification of Mesozoic reptiles, see Benton (2004) and Palmer (1999): earlier classifications are given by Romer (1966) and Young (1981). See also Blount and Crowley (2001). The classification of dinosaurs has been given extensive treatment by Weishampel et al. (1990), Currie and Padian (1997), and Farlow and Brett-Surman (1997), among others.

1.4 Note on Coloration

There is absolutely no way in which the colours of early fossil reptiles can be ascertained but, from what is known about animal coloration today, it is extremely likely that the same principles would have applied then. In all probability the larger dinosaurs would have been uniformly grey, like modern elephants, rhinoceroses and hippopotamuses. Marine reptiles were almost certainly dark above and whitish below, like modern sharks and whales. The effect of a pale ventral surface is that it tends to counteract the effects of shadow and to render its possessor inconspicuous – an invaluable asset to both predators and their potential prey. The same would have applied to flying reptiles as well as to many of the smaller terrestrial forms.

In addition to a general colour resemblance to that of the environment, there may well have been changes in coloration during various stages of the life history. Some extinct reptiles would have been endowed with disruptive coloration which drew attention away from their shape and thus hindered recognition. Others, especially smaller ones, were probably aposematic as are coral snakes today, to advertise distasteful or dangerous characteristics. Alternatively, brightly coloured forms could have been mimics whose models were very conspicuous. Mimicry is very prevalent among coral snakes. Conspicuous colours in adults, especially males, might have had an epigamic or sexual function (Sect. 9.3.1). Indeed, all the functions of animal coloration that we see in the modern world must have been expressed in the colours of Mesozoic reptiles (Halstead and Halstead 1981; Lambert 1992). No attempt, however, has been made in most of the drawings that illustrate this book to guess about the colour patterns of the reptiles depicted. To do so might be more misleading than to depict them as they would have appeared had they been mainly monochromatic or self-coloured.

2 Diversification of Palaeozoic Reptiles

2.1 Introduction

During the Mesozoic Era, the climate of the world was generally warm and moist. Although there were arid times and regions, much of the land surface was in many ways similar to that of the Amazon valley of today, with vast swamps and rainforests. At higher altitudes and away from the swamps there were more open areas with less dense vegetation. The swamp fauna during most of the period consisted of tree ferns and horsetails, that of the open regions of giant conifers, cycads and ferns. As shown in Table 3, of the five major extinctions in the fossil record, one took place at the end of the Permian period, another at the end of the Triassic, and the most recent one at the end of the Cretaceous. Minor extinctions also occurred during the Permian period, and throughout the Mesozoic Era (Fig. 1).

At the end of the Triassic, there was an outburst of volcanic activity in what is now North America, but during the succeeding Jurassic period the land was generally low-lying, with lakes, pools, meandering rivers, and swamps, all liable to invasion by the sea. There was a continuation of mild, subtropical climates and, consequently, the vegetation was similar to but more luxuriant than it had been in the Trias. The early Cretaceous saw a continuation of the Jurassic flora but, by the Middle Cretaceous, a rich and almost modern flora had become widespread. This included beech, poplar, magnolia, laurel, willow, and fig trees. Flowers gave colour to the scene and insects thrived. The climate continued at first to be mild, but later, as continental drift increased, seasonal changes

Table 3. Dates and intensities of the five major extinctions in the fossil record. Based on Jablonski (1989)

	Millions of years before the present	Observed loss of families[a] (%)	Calculated species loss (%)
Ordovician	440	26 (12)	84
Devonian	350	22 (14)	79
Permian	250	51 (52)	95
Triassic	205	22 (12)	79
Cretaceous	45	16 (11)	70

[a] Percentage loss of families of marine invertebrates is shown in brackets

set in. Finally, with the great mountain-building movements of the Laramide revolution (which produced the Rocky Mountains and the Andes), much of the old swamp and lakeland was drained and the 'Age of Reptiles' came to a close. But what had the world been like before it began?

The earliest known fossils of animals that were probably reptiles are among the stem-reptiles (parareptiles). There may well have been a small extinction about this time – between the Mississippian and Pennsylvanian periods (into which the Carboniferous is frequently divided). Only a number of ammonoids and some brachiopods seem to have disappeared, however, which is surprising because a major phase of glaciation began then. As the ice caps grew on Gondwanaland (the southern of the two continents into which the land was then separated), the sea level fell. But there was no mass extinction, either in the sea or on land.

2.2 Continental Drift

In order to interpret the distribution and consequent diversification of the Mesozoic reptiles, one has to realise that continents are by no means static. During Cambrian times, some 500 mya (million years ago), they were distributed along the equatorial belt of the world. They included Laurentia, Siberia, Baltica, Avalonia and the enormous Gondwanaland, the only one to extend into temperate latitudes. During the course of Silurian and Devonian periods, approximately 400 mya (Table 1), Gondwanaland moved southwards while Siberia shifted towards the north, causing major changes in the circulation of the oceans. Later, in the Carboniferous, there was a single supercontinent in the south, Gondwanaland, which comprised what is now South America, Africa, Australia, India and Antarctica, and another, Laurentia, to the north. The latter consisted of North America, Greenland, north-western Europe, and Russia west of the Urals. It is possible that these two vast landmasses may have been connected across the Equator. They became fully united to form an even larger supercontinent, named Pangaea during the Late Triassic and Early Jurassic or Lias, only to split apart again during the Cretaceous period. Finally, with the further break up of both supercontinents, the Atlantic opened up, India began its long migration from Antarctica to Asia, and the earth slowly assumed its present appearance. These 'tectonic' movements, caused by convection currents of molten magma in the earth's mantle (below the solid crust of the planet) were responsible for continental drift, and many of the climatic changes outlined above (Benton 1996).

2.3 Reptilian Ancestors

Although no definitive fossils have been found of the amphibian ancestors from which the reptiles evolved, they are usually considered to have been an early egg-laying offshoot of an amphibian group (Batrachosauria) of which *Seymouria* (Fig. 4) is an example. *Seymouria, Solenodonsaurus* and the larger *Diadectes* (Fig. 4) have sometimes been classified as basal reptiles – but the point at which reptilian characters dominated over those of amphibians is unclear. The remnants of lateral lines on the skull of adult *Seymouria* suggest that this genus had an aquatic development, while *Solenodonsaurus* had a dentition intermediate between that of an amphibian and a reptile. (Lateral lines contain sense organs that respond to pressure and thus enable their owners to detect nearby moving objects.)

Increase in size followed quickly after the reptiles began to radiate on land. For these reasons, and perhaps somewhat arbitrarily, I have included *Diadectes* among the basal reptiles although it is often regarded as a batrachosaur. By the end of the Carboniferous period there were many small captorhinomorph (anapsid) reptiles; while Pelycosauria (Synapsida) measuring more than 1 m in length had also evolved. *Captorhinus* and *Labidosaurus* are among the better known parareptiles from the lowest strata of the Permian. They had high, narrow skulls and pointed noses, which are characteristic of reptiles rather than of amphibians.

The basic physiological character that separates reptiles from amphibians is, unfortunately, not revealed in the fossil record. This is the possession of 'cleidoic' or enclosed eggs. Amphibian eggs develop and hatch in water, but those of reptiles are provided with a semi-impervious shell or membrane, and

Fig. 4. *Above Seymouria* (Batrachosauria; Lower Permian; length ca. 60 cm). *Below Diadectes* ('Parareptilia'; Lower Permian; length ca. 1.5 m)

the presence of water is not necessary for their survival. Ammonia, the chief excretory compound of Amphibia, is extremely soluble in water, but it is toxic when concentrated and would be unsuitable as the nitrogenous waste product of an enclosed embryo. Like insects and birds whose eggs are also laid on land, reptiles eliminate nitrogenous waste in the form of insoluble uric acid. Urea, the principal excretory compound of mammals, would not be suitable as the excretory product of an embryo in an enclosed egg because, when concentrated, it would upset the osmotic relations of the developing embryo. (The urea produced by a mammalian embryo passes through the placenta and is excreted by the mother.) Uricotelic metabolism originally evolved in relation to the possession of cleidoic eggs. It is retained in the developmental and adult stages of reptiles, as it is in insects and birds, because it has the advantage that nitrogenous wastes can be excreted with minimal loss of water.

2.4 Anapsid Reptiles

The early reptiles increased and diversified throughout the Permian period, during which time the world's main reptilian types appeared. Nearly all the individual orders became established during the Triassic, but most of them only reached their maximum development in the Jurassic and Cretaceous periods. We shall now consider some of the earliest forms in greater detail.

Many of the earliest Anapsida, such as *Hylonomus* (Captorhinidae) from the Middle Carboniferous of Canada, were insectivorous. The remains of these animals are well preserved in the ancient tree stumps into which they crawled in pursuit of insects and worms. Although it closely resembled some of its am-

Fig. 5. *Bradysaurus* ('Parareptilia'; Upper Permian; length ca. 2.5 m). (After Colbert 1965)

Fig. 6. *Scutosaurus* ('Parareptilia'; Middle Permian; length ca. 2.5 m). (Based on Špinar 1995)

phibian contemporaries, *Hylonomus* had a high skull – a typical reptilian feature associated with additional jaw muscles – and almost certainly would have laid cleidoic eggs, although none of these has yet been found (Benton 1996). Other early reptiles had several parallel rows of teeth on both the upper and lower jaws, a character taken to excess by the Upper Permian *Moradisaurus* which had more than ten rows. This may have been an adaptation to a diet of tough vegetable matter.

Fig. 7. *Mesosaurus* (Mesosauria; Lower Permian; length ca. 75 cm). (Cloudsley-Thompson 1999)

Most Palaeozoic anapsids were parareptiles. As already mentioned, *Diadectes* (Fig. 4) was one of them. Later forms included Pareiasauridae such as the Late Permian *Bradysaurus* (Fig. 5) a large plant-eater about 2.5 m in length, and the massive Upper Permian *Scutosaurus* (Fig. 6), which was also herbivorous.

Despite the fact that the Testudines or Chelonia had evolved during the Permian, they did not begin to radiate until the Triassic period and will therefore not be discussed until Chapter 4. Another marine order was the short-lived Mesosauria of the Lower Permian period. *Mesosaurus* (Fig. 7), the only known genus, was a highly specialised, long-tailed aquatic form, 75 cm–1 m in length with broad hands and feet. The slender jaws were lined with needle-sharp interlocking teeth well adapted either for filter-feeding or for catching fishes (Sect 4.1).

2.5 Synapsid Reptiles

Primitive synapsids were already present in the Late Pennsylvanian and the group took a leading role in the reptilian radiation of the Permian period. The order Pelycosauria is represented most abundantly in the Upper Carboniferous and Lower Permian strata of North America. Most of the pelycosaurs were small- to medium-sized insectivores and carnivores with strong skulls and sharp teeth. One of these was *Varanosaurus* (Fig. 8), a typical carnivore, about 90 cm in length with a lizard-like appearance little different from that of the anapsids and the primitive diapsids. Others, however, became adapted to a

Fig. 8. *Varanosaurus* (Pelycosauria; Lower Permian; length ca. 90 cm). (Based on Špinar 1995)

Fig. 9. *Edaphosaurus* (Pelycosauria; Upper Carboniferous; length ca. 3 m)

Fig. 10. *Dimetrodon* (Pelycosauria; Lower Permian; length ca. 1.8 m). (Cloudsley-Thompson 1999)

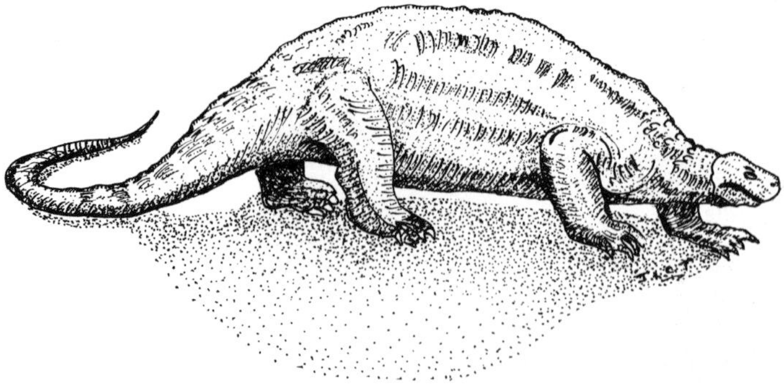

Fig. 11. *Cotylorhynchus* (Pelycosauria; Middle Permian; length ca. 3 m)

vegetarian diet and were among the first terrestrial vertebrates. A few of the later pelycosaurs, such as *Edaphosaurus* (Fig. 9) and *Dimetrodon grandis* (Fig. 10) had large 'sails' supported on vertical spines of the vertebrae. These were probably thermoregulatory devices and acted as radiators (Tracy et al. 1986). In the morning, *D. grandis* would turn sideways to the sun. It would thus absorb heat with its sail, whilst during the hot part of the day it might have faced the sun or found a shaded retreat and radiated excess heat from the sail (see Sect. 7.5.2). Haack (1986) calculated that it would have taken a large *Dimetrodon* (weighing some 250 kg) about 12 h basking in sunshine to increase its body temperature from 25 to 30 °C without a sail, but with its sail it would have taken only 3 h. The ability to warm up comparatively quickly would greatly have benefited it in catching its prey. The weakness of this argument, however, lies in the fact that most pelycosaurs and their other contemporaries lacked sails yet seem to have survived quite well (Benton 2004). A very large pelycosaur, *Cotylorhynchus* (Fig. 11), from the Middle Permian of North America was the ecological equivalent of the gigantic therapsids that lived in South Africa and Russia about the same time.

The second order of Synapida, the Therapsida (Sect. 8.2), encompassed the mammal-like reptiles. The rocks of the Karoo basin of southern Africa are littered with the fossilised bones of these synapsid reptiles. They include both herbivorous and carnivorous Dinocephalia ('giant heads'). Some of the former were quite large. One genus, *Moschops* (Fig. 12), had massive shoulders and neck. Although the limbs and head were comparatively small, the skull was considerably thickened, as in other dinocephalians, and may have been adapted in this manner for head-butting contests at the time of mating. These thick-skinned reptiles formed the prey of carnivorous forms such as the small *Titanosuchus* (Dinocephalia; Fig. 13) and *Sauroctonus* (Gorgonopsia; Fig. 14) a large sabre-toothed therapsid. Although the former was a slender hunter that probably fed mainly on smaller herbivores and juveniles, many of the other titanosuchids were large, heavily-built predators and could not have been very agile or fast. The most common of the herbivorous therapsids were the Dicy-

Fig. 12. *Moschops* (Therapsida; Middle Permian; length ca. 1.8 m). (Based on Špinar 1995)

nodontia ('two dog teeth'). These had large, sausage-shaped bodies, short legs and tails. They either possessed no teeth at all, or had just a pair of tusks; and they sliced the vegetation with sharp, horny beaks. A complex chewing cycle allowed them to exploit a wide variety of plants. When feeding, the lower jaw was first thrust forward, slid back forcibly cutting the food, and then opened from a backwards position. They were preyed on by gorgonopsians along with the carnivorous and other plant-eating dinocephalians. Neither dinocephalians nor

Fig. 13. *Titanosuchus* (Therapsida; Middle Permian; length ca. 1 m)

Fig. 14. *Sauroctonus* (Therapsida; Upper Permian; length ca. 2.5 m). (Based on Špinar 1995)

gorgonopsians survived after the Permian, but the dicynodonts flourished for a second time during the Triassic period.

The mass extinction at the end of the Permian period (ca. 250 mya) was the largest of all time (Table 3; Sect. 12.1.2). It has been estimated that considerably more than half the families of animals that had existed during the Late Permian abruptly disappeared. These included numerous groups of marine invertebrates such as the trilobites, many kinds of coral, and most of the brachiopods. Molluscs and fishes also suffered severely. The effects were equally devastating on land. Most of the primitive amphibians and of the mammal-like reptiles vanished (Jablonski 1989; Hallam and Wignall 1997; Benton 2003). Although surviving groups of the latter were eventually to give rise to the mammals (which almost certainly evolved from dicynodonts) the predominant reptiles of the Mesozoic Era were Diapsida. The former included the marine ichthyosaurs and plesiosaurs: we shall deal with them before considering aerial and terrestrial reptiles. Benton (1996) has given a very clear account of continental drift and of conditions during the Carboniferous and Permian periods.

3 The Mesozoic Environment

3.1 Introduction

The fossils of the first tetrapods, early amphibians (Chap. 1), have been found in Upper Devonian strata of the eastern part of Greenland. The flora at that time was surprisingly varied: it included dense growths of increasingly tall club mosses, horsetails, and ferns. The Carboniferous (350 mya) and Permian (270 mya) periods were both warm in the northern hemisphere. In the south, however, a large ice cap developed which covered considerable portions of Gondwanaland (Benton 1996, 2004).

Until the Carboniferous period, land plants had been restricted to moist areas around rivers, ponds and lakes: they were by no means large. During the Carboniferous, however, when reptiles first appeared, giant horsetails, ferns, and seed ferns or pteridosperms evolved. Seed ferns resembled true ferns (pteridophytes) in many ways, but had already evolved seeds. They bridged the gulf between ferns and cycads. The latter are palm-like plants and the pteridosperms were more closely related to them than to pteridophytes.

Ferns, tree ferns – which reached to a height of 9–15 m – horsetails and giant club mosses were characteristic of the Carboniferous period. One genus of the club mosses, *Sigillaria*, grew to ca. 20 m and was one of the main components of the coal forests of the northern hemisphere. Its cones grew directly from the trunk, just below the apex. Spore-bearing ferns predominated in the Early Permian, but by the end of the period they had been largely replaced by gymnosperms (coniferous trees). This change in the vegetation during the Permian was very pronounced, and marked a transition from the early to the middle phase of plant development. Giant club mosses gradually disappeared, along with many pteridosperms and some of the horsetails, while new types of ferns and ginkgo trees appeared as the climate became drier. In Gondwanaland, ferns of the genus *Glossopteris,* with long, tongue-shaped leaves and central veins, as well as of *Gangamopteris*, with leaves from which the central veins were absent, were characteristic of this time.

3.2
The Triassic Period

The Triassic period (250–205 mya) during which the dinosaurs originated, was a time of comparative aridity, of xerophytic vegetation and abundant reptiles. About 200 mya, during the Late Triassic and Early Jurassic, the continental plates of Laurasia and Gondwanaland were united. Some portions of what are now central Asia and China may have remained as separate islands, but most of the landmass of the world was concentrated into the single supercontinent, Pangaea, as we have seen. Consequently, both plants and animals were remarkably similar everywhere. The main vegetation types consisted of seed ferns and conifers which were adapted to the dry climate, while massive horsetails dominated the damper regions. The terrestrial fauna comprised mainly earthworms, again in moister places, snails, arachnids, insects, and reptiles, which either captured earthworms and arthropods in the undergrowth, or else chewed the tough vegetation. Crustaceans, molluscs and fishes populated the lakes and rivers where they were preyed upon by amphibians and phytosaurs – distant relatives of the crocodilians (Sect. 4.6.2). Piscivorous ichthyosaurs and nothosaurs dominated the oceans, while early pterosaurs skimmed squids and fishes from the surface water. Numerous marine taxa became extinct at the end of the Triassic, including the primitive archosaurs – the phytosaurs, aetosaurs, and rauisuchians. These extraordinary animals will be discussed in the following chapters. The dinosaurs evolved from thecodont reptiles, and the mammals from theropsids about 230 mya.

3.3
The Jurassic and Cretaceous Periods

In contrast to the Triassic, the Jurassic period (205–145 mya) was a time of moist climates and luxurious forests. The first flowering of giant dinosaurs began towards the end of the Early Jurassic, and a wealth of dinosaurian fossils has been found in Late Jurassic strata. Indeed, the Late Jurassic was the time during which the largest of the dinosaurs existed.

In many ways, the climates of the Cretaceous (145–65 mya) resembled those of the Jurassic, but towards the end of the period the world became drier and, as already mentioned, the climate seasonal. During the Early Cretaceous, the sea level was some 25 m higher than it is today. About 100 mya, however, it rose dramatically to about 200 m above the present level. All the above-mentioned factors may have contributed to the extinctions that occurred at the end of the Mesozoic Era (Chap. 12). Unlike the Triassic and Jurassic periods, which are both divided into three epochs, the Cretaceous is separated into two.

The predominant elements of the Mesozoic flora were gymnosperms (Fig. 15). Among them were various Cycadoidea – low forms with short, globular or barrel-shaped trunks sometimes bearing large and brightly coloured flowers. There were also tree-like species with slender stems that branched only occasionally.

Fig. 15. The Mesozoic landscape

Both types bore rich crowns of long, palm-like leaves. When the conifers appeared they spread rapidly and soon became abundant. Alongside these were various types of ginkgo, pine-like taxodiums, and huge sequoias. Cryptograms lost their former leading position among the vegetation. Ferns and horsetails still grew in damp places along the banks of rivers or lakes, but they were much smaller than they had been previously. In the absence of grass, the ground was either bare or covered by algae, bryophytes, and pteridophytes. Finally, during the Cretaceous period, as the seasons set in, deciduous plants appeared. These shed their leaves during the dry season or in winter, when climatic conditions became unfavourable, and thus the vegetation came to resemble that of the world today (Špinar 1995).

The vegetation during the Mesozoic Era was not only different from that of earlier times, but was also more diverse. Consequently the fauna diversified also to fill the numerous ecological niches that appeared. This diversity is reflected among the wonderful reptiles discussed in the following chapters.

4 Amphibious and Early Marine Mesozoic Reptiles

4.1 Introduction

Although it might seem logical to discuss the dominant land reptiles of the Mesozoic Era immediately after considering the diversification on land of Palaeozoic reptiles, it is more convenient first to dismiss those reptilian taxa that swam in the seas and flew in the air. Colbert (1966) followed the same procedure, as did Palmer (1999) amongst others. Land-dwelling vertebrates did not turn to life in the sea until the Triassic period, but this trend has been repeated several times since then. Many amphibians were aquatic in the Palaeozoic Era and in the Triassic period, but few of them were marine. The reason for this probably lies mainly in their osmotic relations. Reptiles were better equipped to evolve the adaptations necessary for a marine existence. Not only did they have efficient lungs which had replaced their long-lost ancestral gills, but their limbs were easily transformed into paddles. These, like their tails, in some cases, acquired functions very similar to those of the fins and tails of fishes. Furthermore, reptiles had cleidoic eggs (Sect. 2.3): some came ashore to bury them in the sand, but others evolved ovoviviparity – the birth of living young which had developed from eggs in the body of the mother where they were relatively safe.

The factors that most influence the ways in which animals have evolved and specialised are, first, the environment in which they live and move and, second, what they feed on. Tetrapod reptiles evolved and first diversified on land. Some of them then returned to water and became adapted for swimming; others developed wings and took to the air like birds. (The theropod origin of the latter forms the subject matter of Sect. 11.5.) The majority of reptiles, however, remained on land. These will be discussed in Chapters 7–11. The above is, of course, a very great over-simplification: between such broad groupings are many intermediate forms. Some of the land-living reptiles were shore-dwelling but semi-aquatic or amphibious forms that spent some parts of the day on land and some in water as modern crocodilians do. They moved reasonably well on land (Sect. 4.6), yet were equally at home in water where they swam, often with webbed hands and feet or by means of flattened tails.

Not only were many Mesozoic reptiles amphibious, but several became truly aquatic. These included some of the turtles (Testudines or Chelonia), the placodonts, nothosaurs, and plesiosaurs ('near lizards'; Sauropterygia), sea crocodiles (geosaurs and thalattosuchians: Archosauria) and the Ichthyosauria, which were the most dramatically adapted of them all. A total of seven subclasses or orders – either entirely or in a large portion – became aquatic and returned to the waters in which their distant rhipidistian ancestors originally evolved. One of these was the family Mesosauridae, an example of which, *Mesosaurus* (Fig. 7) from the Lower Permian of southern Africa and Brazil, has already been mentioned in this book (Sect. 2.4). Another mesosaur, *Stereosternum*, has only been found in southern Brazil. Were any needed, this distribution provides unequivocal evidence for continental drift. Although the mesosaurs have sometimes been regarded as suspension feeders, this reconstruction depends upon interpretation of the mandibular teeth as being small, marginal upper teeth. According to Špinar (1995), as well as Collin and Janis (1997), it is therefore more probable that the mesosaurs were piscivorous, living in fresh water and capturing fish and other small individual prey items selectively, rather than by filtering large volumes of water through their teeth – but neither hypothesis is yet fully substantiated. Benton (2004), among others, has favoured the latter, Collin and Janis (1997) the former.

4.2 Locomotion

Many aquatic reptiles use their limbs as paddles and swim relatively slowly. Turtles almost fly through the water with their front flippers but, even so, they are not as fast as most fishes and dolphins of comparative size. Some reptiles, however, especially the pliosaurs and ichthyosaurs, were streamlined and extremely fast (Chap. 5). They swam with the aid of flattened bodies and fish-like tails. When an eel swims, the inner sides of the curves on its sinuous body press against the water. The resistance is increased by the expansion of the dorsal, caudal and ventral fins which are combined into one.

The machinery necessary for typical swimming has been analysed by Sir James Gray and Hans W. Lissman (Gray 1968) and consists of three parts – the vertebral column, the axial musculature and the lateral surfaces of the body, including the caudal fins. The vertebral column can be regarded as a series of rigid units, hinged so that the body can bend only in a transverse lateral plane. The whole tail is essentially a shortened continuation of such units. Eel-like or anguilliform movement is similar to the serpentine locomotion of a snake. Its dependence upon the thrust made by the inner side of each curve of the body can be illustrated by the inability of an eel or snake to crawl across a polished surface. Both, however, can pass through a series of smooth pegs projecting from the surface of a polished board (Fig. 16). The sinusoidal curves of a snake's body when it is thus gliding by serpentine motion are stationary relative to the ground. The same is true of the swimming eel. It is the animal's body that moves, each part closely following the path of the part in front.

In most fishes, and other animals whose bodies are relatively much shorter than those of eels and snakes, the lateral surfaces and caudal fins represent

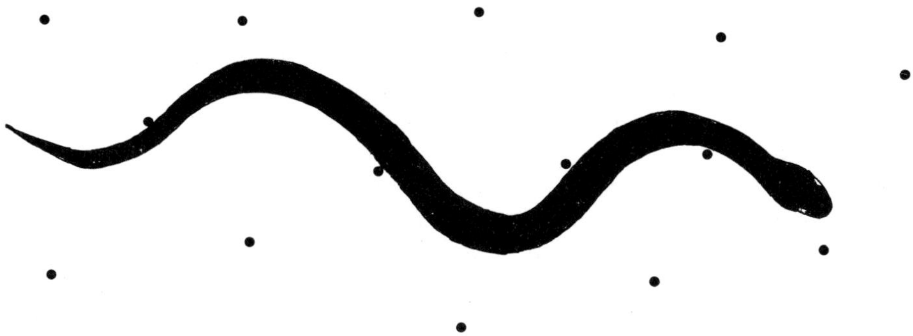

Fig. 16. Snake gliding past a series of smooth pegs projecting from the surface of a polished board. (Cloudsley-Thompson 1999). Further explanation in the text

only sections of the sinuous curve. Furthermore, the separated dorsal, ventral and caudal fins then often have more than one function (Gray 1968; McGowan 1991). When snakes swim, much of the energy is wasted because their rounded bodies present little resistance to the water. Even in sea snakes, the tails are only slightly flattened (see Cloudsley-Thompson 1999). The mechanics of swimming have been analysed very thoroughly by McGowan (1991).

According to Massare (1997), marine reptiles past and present can be divided into four groups according to their body forms or Baupläne (Fig. 17). Each of these was characterised by a different mode of swimming. The first group (Bauplan I) consisted of post-Triassic and some Triassic ichthyosaurs (Chap. 5).

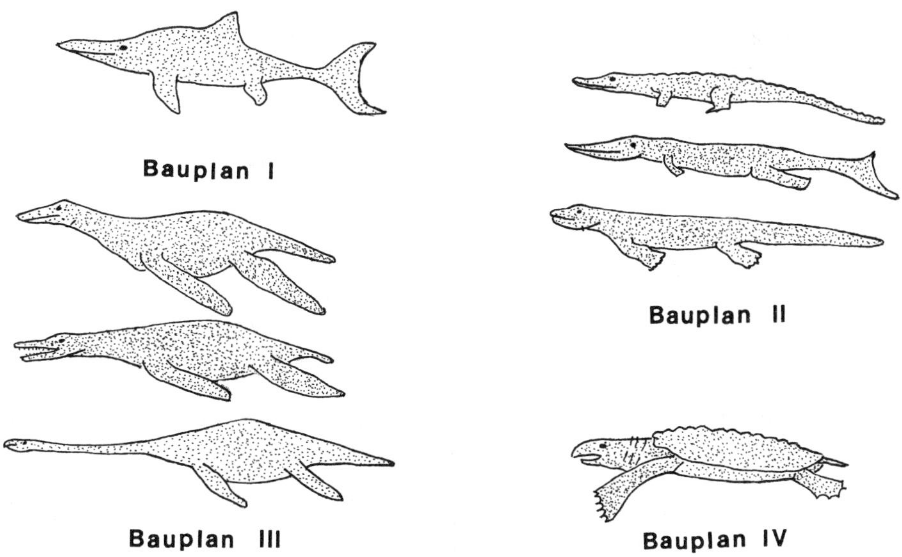

Fig. 17. The four body forms or Baupläne displayed by Mesozoic reptiles. (After Massare 1997)

These had deep, streamlined bodies, deepest in the pectoral region and tapering posteriorly to the caudal fin. They propelled themselves by oscillating the tail, while the fin-like limbs were used mainly for steering. The general shape of the body was within the optimal range to minimise drag and for efficient swimming. Probably the fastest marine reptiles capable of sustained swimming, they were predators that hunted fishes and other prey over large areas.

Bauplan II contained the mosasaurs, thalattosaurs, marine crocodilians, pachypleurosaurs, and the remaining Triassic ichthyosaurs. These animals had narrow, elongated bodies and long, broad tails. Propulsion was achieved by undulation of part or all of the body along with the tail. Consequently, the tail could be used to produce a sudden thrust, and the prey must have been caught by rapid bursts of speed.

Bauplan III is exemplified by nothosaurs, pliosaurs and plesiosaurs; these had stiff, ellipsoidal bodies with two pairs of elongated limbs, wing-shaped in the plesiosaurs. Swimming consisted of a modified underwater 'flight' somewhat similar to that found among extant sea lions (Fig. 32; Chap. 5). The nothosaurs and plesiosaurs had long necks and small heads, while the pliosaurs had shorter necks, larger heads and more compact bodies.

Although not so speedy as the ichthyosaurs, but like them, the pliosaurs were probably predators that pursued their prey over long distances. In contrast, the other taxa would have been much slower. Movement of their necks and heads must have tended to cause them to veer off course if they moved too fast; so they ambushed their prey by stealth and seized it with a sudden movement of the neck.

Finally, Bauplan IV, found in placodonts and sea turtles, was a body compressed dorsoventrally and covered by bony armour. Propulsion was achieved by paddling with the limbs, although in some turtles the forelimbs were modified into 'wings', and propulsion was by subaqueous flight, thus achieving greater speed The placodonts, which lacked such adaptations, were incapable of speed or acceleration and preyed on immobile shellfish. Locomotion was therefore not an important factor to them in the acquisition of food.

4.3 Testudines

The Harvard palaeontologist, A.S. Romer (1966), once remarked that tortoises, terrapins and turtles are commonplace objects to us only because they are still living. Were they extinct, their shells would have been a cause of wonder, as they represent the most complete defensive armour found among tetrapod vertebrates. The Testudines or Chelonia is an ancient order of reptiles long considered to belong to the subclass Anapsida and therefore descended from the Captorhinidae or cotylosaurs discussed in Chapter 2 (Sect. 2.4). Turtles and their relations are the only living descendants of this varied taxon, and there are no other anapsid reptiles alive today. Chelonians are probably descended from the Permian Eunotosauria, and true turtles ap-

Fig. 18. *Left Proganochelys* (Testudines; Upper Triassic; length ca. 70 cm). *Right Triassochelys* (Testudines; Upper Triassic; length ca. 70 cm)

peared during the Middle and Late Triassic (some 220 mya) along with the early dinosaurs.

Modern turtles have no teeth, but the Triassic chelonians *Proganochelys* and *Triassochelys* (Fig. 18) still had a few teeth on their palates. So, too, had the Australochelidae from the Late Triassic of South America and the Early Jurassic of Africa, which still retained some palate teeth as in *Australochelys.* Teeth had disappeared almost completely by the time the Jurassic forms evolved.

4.3.1 Respiration

On account of their rigid shells, respiration presents special problems for Testudines, because it is not possible to pump air in and out of the lungs by movements of the chest. Furthermore, the lungs are emptied when those species that are able to do so withdraw their heads and limbs into the shell. A turtle or tortoise that has been attacked or disturbed may have to remain in this state for a long period. It is able to do so, however, thanks to the evolution of a number of physiological mechanisms which must already have been present in its Mesozoic forebears. These include the ability to tolerate high levels of carbon dioxide in the blood. In other vertebrates these increase the acidity (or decrease the pH), but in chelonians this effect is counteracted by means of a buffering system of bicarbonate ions, serum proteins, and haemoglobin in the blood. In addition, turtles are capable of filling and emptying their lungs more completely than most other vertebrates can. In consequence, when the head and limbs are withdrawn, the lungs initially contain very little carbon dioxide.

The muscles of turtles are often surprisingly red, due to their high myoglobin content. These animals are therefore able to store significant amounts of oxygen, not only in the haemoglobin of the blood but also in the myoglobin of the muscles. In addition, they can incur an 'oxygen debt', which is later paid off by panting or breathing rapidly. In setting up an oxygen debt, energy is supplied to the muscles by the anaerobic conversion of ionised pyruvic acid, with the production of lactic acid, which is stored in the muscles and later conveyed to the liver to be oxidised. Although it is somewhat toxic, turtles can tolerate much higher levels of lactic acid than most other animals. A modern land tortoise can voluntarily holds its breath for periods of 15 or 20 min and may, in this way, economise in water (Cloudsley-Thompson 1999). The ability to spend long periods without respiring is useful, not only to terrestrial chelonians that retract their extremities when disturbed, but also to aquatic species, enabling them to remain submerged for long periods. These periods are extended by bradycardia (slowing of the heart rate), which occurs automatically when aquatic turtles dive. Furthermore, these reptiles are able to extract dissolved oxygen from the surrounding water through their buccal linings or thin papillose skin (Pritchard 1979).

Movement of air in and out of the lungs is achieved in a number of ways. The pressure in the lungs is continuously being changed by the heartbeat and the movements of the limb girdles, limbs and neck. A membrane behind the lungs can also compress these organs causing exhalation. In marine turtles whose limbs are not retractable, some expansion of the lungs is possible because there is skeletal flexibility of the bridge between the carapace and plastron.

4.3.2 Evolution

Chelonians have changed very little since they first evolved, yet they have survived all the natural catastrophes that eliminated so many other reptilian taxa during the Mesozoic. No doubt their life styles also have not altered much during the Era; but *Proganochelys* and *Triassochelys* (Proganochelyidae; Fig. 18) may not have developed the ability to retract the head, legs and tail within the shell as completely as the more advanced modern chelonians do. Both these genera were found in the Upper Triassic Keuper sediments of Europe. They were probably semi-aquatic, like modern terrapins: marine turtles did not appear until the Jurassic period. Could that ecological niche have been already occupied by the placodonts (Sect. 4.4.1)? Undoubtedly, tortoises and turtles on the shore would, in Late Triassic times, have seen nothosaurs, placodonts, plesiosaurs and ichthyosaurs in the waters beyond the beaches (Colbert 1965).

From the Jurassic to the present day, Testudines have comprised two major groups, the Pleurodira and Cryptodira. The former withdraw their heads by bending their necks sideways, while the cryptodires bend them vertically. Although living pleurodires, such as the matamata (*Chelus fimbriatus*) and other snake necks (Chelyidae), are all freshwater animals of southern continents, fossil species have been found in all parts of the world and may include marine as well as freshwater forms.

Fig. 19. *Archelon* (Testudines; Upper Cretaceous; length ca. 3.7 m)

Mesozoic reptiles are renowned for their large size, and the Testudines were no exception. The giant turtle *Archelon ischyrios* (Cryptodira; Fig. 19) from the Upper Cretaceous of North America measured over 4 m in length and, even though its shell was reduced to a framework of trombone-like ribs, may have weighed about 3 tonnes. These ribs were probably covered by a thick coat of rubbery skin, as is the extant leatherback turtle (*Dermochelys coriacea*), rather than by the usual horny plates. There were fewer predators in the sea than on land in those days, so the greater manoeuverability afforded by reduction in weight would have offset the defensive advantages of heavy armour on the back. Furthermore, the limbs were powerful with extremely long, webbed digits that formed broad flippers to propel the animal through the water. *D. coriacea*, the largest living turtle, measures up to 2.1 m from the tip of the beak to the end of the tail (carapace length 1.5 m) and weighs as much as 4.5 kg. No marine turtle, either extinct or extant, has been able to retract its head or limbs into the shell as terrestrial chelonians do.

4.4 Placodonts and Nothosaurs

4.4.1 Placodonts

In Lower Triassic times, the Placodontia ('plate teeth') became specialised for life in shallow water. These slow, massive reptiles had stout bodies, short necks and tails, and paddle-like limbs. In some genera, such as the remarkable *Heno-*

Fig. 20. *Henodus* (Placodontia; Lower Triassic; length ca. 1 m). (Cloudsley-Thompson 1999 after Špinar 1995)

dus (Fig. 20), the body was very broad and armoured with bony plates, giving an appearance strikingly similar to that of the marine turtles, although they were in no way related to them. Their large, flattened plate-like teeth were obviously adapted for crushing shellfish, and placodonts probably dived to the bottom of the shallow waters to dig up brachiopods, clams, and other molluscs. In some genera, including *Henodus*, the teeth have disappeared altogether, as we have seen and, as in modern chelonians, the jaws were probably covered with horny plates. Ellis (2003) suggested that grooves in the jaws might have held

Fig. 21. *Placodus* (Placodontia; Lower Triassic; length ca. 2.5 m)

Fig. 22. *Left* Head of *Placodus; right* head of *Placochelys.* (Based on Špinar 1995)

baleen-like plates for filter feeding; but what *Henodus* actually fed on is still a mystery. Like the Mesosauria and all other aquatic reptiles, the Placodontia were secondarily aquatic. They were descended from the Protorosauria ('dawn lizards'). The latter were an odd collection of possibly unrelated Permian and Triassic reptiles, some lizard-like in appearance, others with small limbs, elongated bodies, and extremely long necks. The fossil of *Henodus* (Fig. 20) was found near Tübingen in southern Germany. It was one of the last of the placodonts.

The skull of the placodonts was short and the nostrils had retreated to a position immediately anterior to the eyes. The front teeth of *Placodus* (Figs. 21, 22) protruded almost horizontally, and were evidently specialised for plucking shellfish from the seafloor. These were then crushed by the blunt grinding teeth, brought together by strong jaw muscles. As can be seen in Fig. 21, *Placodus* was relatively slim, but the body of *Placochelys* was broad, and heavily armoured like that of *Henodus*, with dorsal scales that coalesced to form a heavy carapace over the back. It had a tapering, toothless snout that may have been used like forceps, and fewer grinding teeth than *Placodus* (Fig. 22; Špinar 1995).

4.4.2 Nothosaurs

The Nothosauria was another suborder of Sauropterygia whose fossils have been uncovered in Europe, Asia, and America. Smaller, elongated marine reptiles, they existed contemporaneously with the placodonts and early ichthyosaurs, reaching the peak of their development in Late Triassic times. Their paddle-shaped limbs and feet were strong, their necks long and flexible, and their nostrils had moved slightly backwards from their snouts. They were probably ecological analogues of today's sea lions and seals, and fed mainly on fish. As

Fig. 23. *Nothosaurus* (Nothosauria; Upper Triassic) length ca. 3 m

they swam slowly along, they would have darted their long, flexible necks from side to side, catching with their numerous sharp teeth any fishes that came within reach. *Nothosaurus* (Fig. 23) was a typical example. The nothosaurs probably spent part of their time fishing at sea, part resting on land. Their feet had five long toes that were webbed to varying degrees – as in *Lariosaurus, Ceresiosaurus*, and *Pistosaurus* (Fig. 24) – and their tails quite possibly bore fins.

Lariosaurus was one of the smaller nothosaurs, but by no means the smallest, for some measured only 20 cm in length. Its short neck and toes were primitive features. The small webs between its toes would not have been of much use in

Fig. 24. *Above Lariosaurus* (Nothosauria; Middle Triassic; length ca. 60 cm). *Centre Ceresiosaurus* (Nothosauria; Middle Triassic; length ca. 4 m). *Below Pistosaurus* (Nothosauria; Middle Triassic; length ca. 3 m). (After Palmer 1999)

swimming. Indeed, *Lariosaurus* probably spent much of its time paddling around in coastal shallows, feeding on small fishes, crustaceans, and molluscs. In contrast, the toes of *Ceresiosaurus* were relatively much longer than those of most other nothosaurs. In addition, the animal exhibited hyperphalangy – the phenomenon in which the number of phalanges in each toe is increased. The flippers would therefore have been efficient swimming organs, anticipating the great oar-like limbs of the plesiosaurs of the Jurassic period (Palmer 1999). Indeed, the nothosaurs may well have been ancestral to the Plesiosauria (Sect. 5.1).

4.4.3 Claudiosaurians

Before we leave the nothosaurs, it may be appropriate to mention *Claudiosaurus* (Fig. 25), a genus of amphibious reptiles that evolved in the Late Permian and persisted into Early Triassic times and whose affinities are at present uncertain. The claudiosaurians were diapsids and, like the nothosaurs, have been regarded as being ancestral to the plesiosaurs. They have also been proposed as a sister taxon to them. In the absence of information on groups specifically ancestral to the thalattosaurs, plesiosaurs, and ichthyosaurs, their origin remains a mystery (Carroll 1997).

Claudiosaurus was a long-necked, lizard-like marine reptile whose lifestyle may well have been similar to that of the marine iguana (*Amblyrhynchus cristatus*) of the Galapagos Islands. It probably fed under water between episodes of basking in the sun, poking its long, flexible neck and small head into fronds of seaweed to find edible plants and animals. Its feet were not webbed, and its legs would have been folded against the body when it swam. Propulsion must have been achieved by undulations of the body and long, narrow tail. Buoyancy would have been enhanced by the large amount of cartilage present in the skeleton. The sternum was not well developed, nor was it ossified as in true terrestrial animals: this indicates that *Claudiosaurus* was not well adapted for locomotion on land (Palmer 1999).

Fig. 25. *Claudiosaurus* (order uncertain; Permian-Triassic; length ca. 60 cm)

4.5
Thalattosaurs, Mosasaurs and Other Sea Lizards

Lizards and snakes (Squamata) and *Sphenodon* (Rhynchocephalia), the lizard-like tuatara of New Zealand, are all classified in the subclass Lepidosauria (Table 2) which originated in the Permian. The first known sphenodontians (Sect. 8.4.1) appeared in the Triassic and had long slender limbs, but *Pleurosaurus* (Fig. 26) from the Late Jurassic or Malm of Germany, and some of the Early Cretaceous forms, were aquatic. Their limbs were reduced while the tail was even longer than the elongated body and probably used for propulsion – there were up to 57 vertebrae in some genera. The nostrils opened back on the snout near the eyes, another adaptation to life in water. True lizards (Squamata) were represented by mosasaurs such as the well-known North American *Tylosaurus* (Fig. 27) and the thalattosaurs. *Thalattosaurus* from the Upper Triassic of California was somewhat smaller than *Tylosaurus*, but both had paddle-like limbs and elongated skulls. As with many aquatic animals, the neck and body were long and slim in *Thalattosaurus*, the tail elongated so that the animal would have swum in an eel-like fashion, using its feet mainly for steering and braking. They were also used on land when the animals came ashore to lay their eggs. *Askeptosaurus* (Fig. 28) had long jaws and many sharp teeth – good for catching fish (Sect. 4.6.4). Its eyes were large and probably adapted for vision in the twilight zone of the ocean. They were strengthened by a ring of bony plates that would have resisted the pressure of the water when the animal dived deeply after its prey. Other varanid-like sea lizards of the Mesozoic were

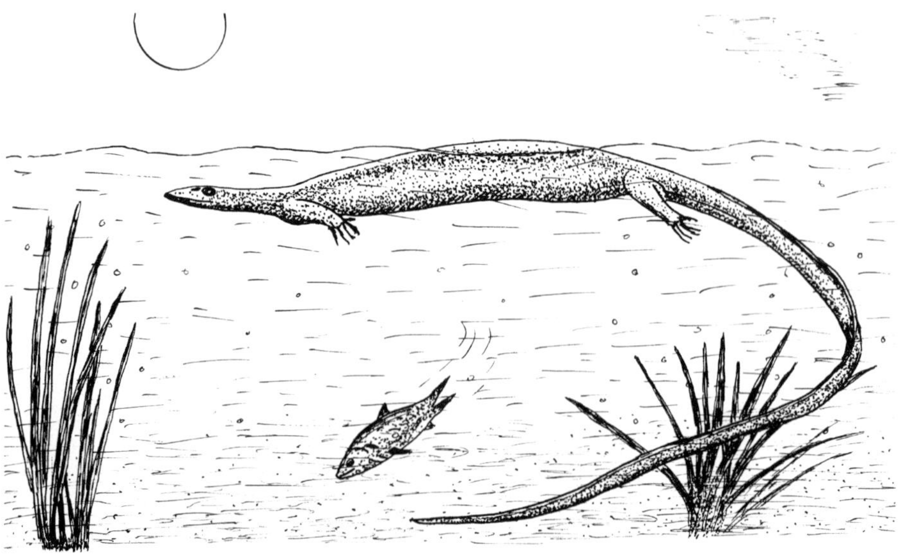

Fig. 26. *Pleurosaurus* (Rhynchocephalia; Upper Jurassic; length ca. 60 cm). (After Palmer 1999)

Fig. 27. *Tylosaurus* (Mosasauria; Upper Cretaceous; length ca. 9 m). (Cloudsley-Thompson 1999)

the Aigialosauria and Dolichosauria. These appeared in Early Cretaceous times and were small animals, only partly adapted for marine life. Indeed, the aigialosaurs appear to have been ancestral to the Upper Cretaceous mosasaurs, which were among the most highly specialised marine reptiles. As time passed, the Mosasauridae, like many other marine vertebrates, gradually became larger and larger: *Tylosaurus* reached over 9 m in length. Their adaptations included a retreat of the nostrils to a posterior position on the top of the skull as in *Pleurosaurus*, modification of the limbs, and broadening of the tail. There

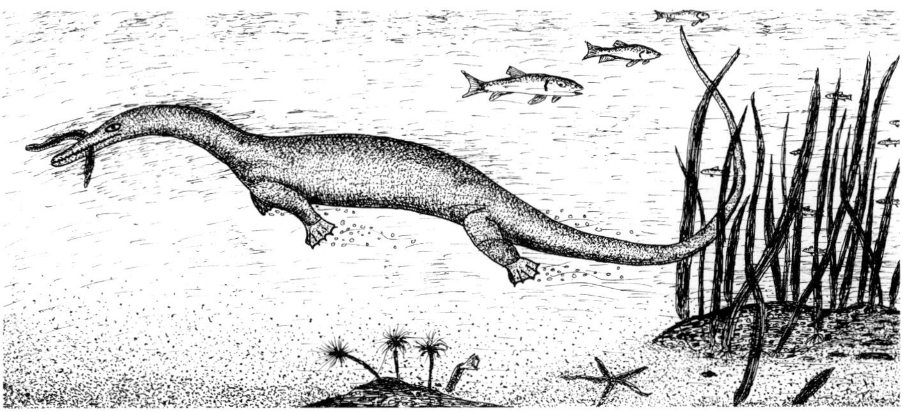

Fig. 28. *Askeptosaurus* (Thalattosauria; Middle Triassic; length ca. 2 m). (After Palmer 1999)

was no bending of the backbone either up or down, the tail extending in side view in a straight line from the pelvic region to the tip (Fig. 26). Clearly, the mosasaurs were highly efficient swimmers, propelling themselves through the water by lateral undulations of the elongated body and flattened tail, and using their limbs for balancing and steering (Colbert 1966).

The Mosasauridae were among the most rapacious predators of the Upper Cretaceous seas. Their jaws were very long and beset with rows of strong conical teeth. The centre of the lower jaw was hinged, which made the gape very elastic so that large chunks of food could be swallowed. No fossils of young mosasaurs have been found, which suggests that the females may have left the oceans and ascended rivers to breed. The young could then have spent their early lives in that more sheltered environment before migrating back to the ocean. Like modern varanid lizards and in common with many other reptiles both living and extinct, the mosasaurs probably used their long, whip-like tails to great effect in defence.

4.6 Champsosaurs, Phytosaurs and Crocodilians

4.6.1 Choristodera

During the late Permian period, two lineages of diapsid reptiles rose to prominence, the Lepidosauria and the Archosauria (see below). Among the former were the Choristodera, a strange assembly of crocodile-like reptiles that diverged from the main diapsid line during the Lower Cretaceous period, some 140 mya. One of the genera included was *Champsosaurus* (Fig. 29). This arose

Fig. 29. *Champsosaurus* (Choristodera; Upper Cretaceous; length ca. 1.5 m). (After Palmer 1999)

in the Late Cretaceous period, persisted into the Eocene epoch, and became extinct only about 50 mya. *Champsosaurus* lived in the rivers and swamps of Europe and North America. It had long, narrow jaws with pointed teeth – like those of the modern gharial (*Gavialis gangeticus*). Although the Lepidosauria and Archosauria were and are both diapsids, they are not closely related. The similarity between *Gavialis* and *Champsosaurus* is due to convergence. Narrow jaws are an adaptation for eating fishes. These are caught with a sideways sweep of the head through the water: it would not be possible to move a broad snout quite so rapidly. *Champsosaurus* probably swam as crocodilians do, by means of lateral undulations of the body and tail, while their limbs were held tightly against the side of the body where they did not interfere with their owner's streamlined shape. The great width of the skull behind the animal's eyes provided a large area for the attachment of strong law muscles, and *Champsosaurus* was capable of giving a very powerful bite. It probably ate birds and mammals as well as fishes, as does the Indian gharial.

4.6.2 Phytosaurs

During the Late Triassic, one line of Thecodontia (Table 2), the phytosaurs, abandoned a bipedal habit on land and took to the water. These creatures resembled crocodilians very closely in structure and, presumably, in habits – but were not ancestral to them. Nevertheless, phytosaur genera such as *Parasuchus, Rutiodon* and *Phytosaurus* (Fig. 30) – with flattened tails, elongated

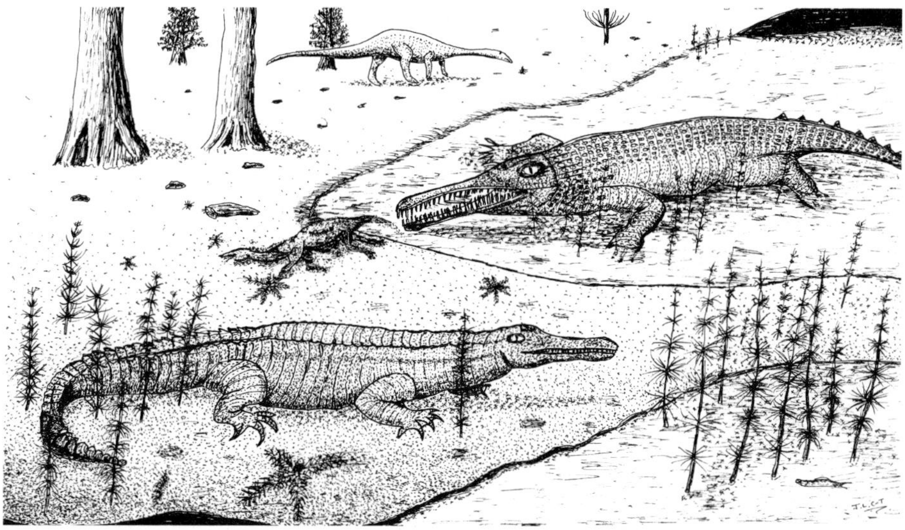

Fig. 30. *Left Rutiodon* (Thecodontia; Upper Triassic; length ca. 3 m). (After Palmer 1999). *Right Phytosaurus* (Thecodontia; Upper Triassic; length ca. 10 m). (After Colbert 1966)

jaws, and rows of sharp conical teeth – showed remarkable parallel evolution with crocodiles. Like those of crocodiles, their hind limbs were longer than the fore limbs; but their nostrils were elevated on a bony lump near their eyes, far back in their skulls. (In crocodiles the nostrils are placed at the tip of the snout.) Furthermore, the phytosaurs had no palate. As in crocodiles, the forms illustrated in Fig. 30 have eyes with slit pupils (Sect. 4.3.4). *Phytosaurus* (as reconstructed by M.M. Colbert) is shown about to scavenge on the decaying carcass of a dinosaur. All the thecodonts, including the phytosaurs, became extinct at the end of the Triassic period but the ecological niche they had once occupied was eventually exploited by true crocodiles.

4.6.3 Crocodilians

The seas and fresh waters of the Jurassic and Cretaceous periods teemed with fishes and other animals preyed on by a variety of reptiles including crocodilians. The earliest of the latter were Sphenosuchia, of which *Gracilisuchus* (Fig. 31a) appeared in the Middle Triassic of South America. This tiny creature

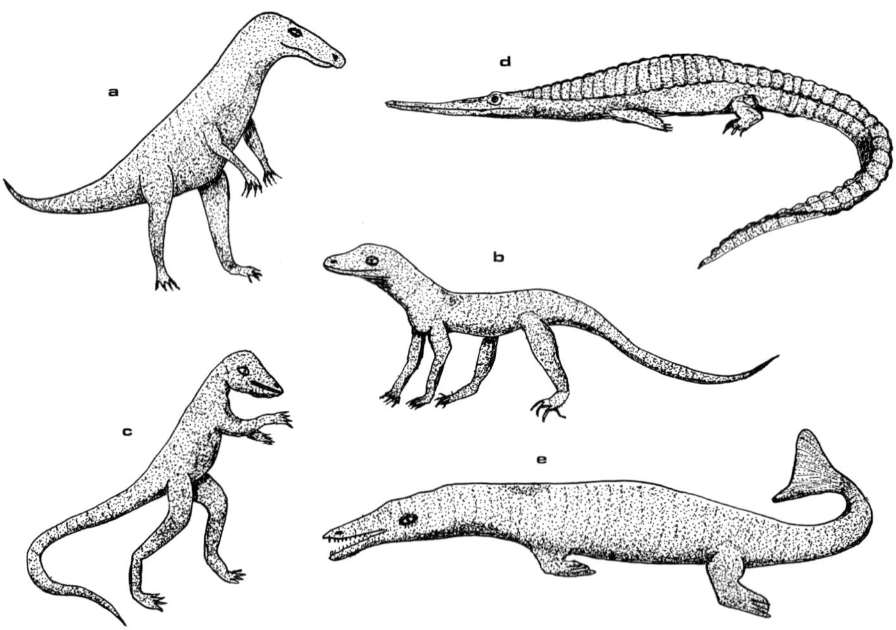

Fig. 31a–e. Early crocodilians. **a** *Gracilisuchus* (Sphenoschidae; Middle Triassic; length ca. 30 cm), **b** *Terrestrisuchus* (Saltoposuchidae; Upper Triassic; length ca. 50 cm), **c** *Saltoposuchus* (Saltoposuchidae; Upper Triassic; length ca. 1.2 m), **d** *Teleosaurus* (Crocodylia; Lower Jurassic or Lias; length ca. 3 m), **e** *Metriorhynchus* (Crocodylia; Middle Jurassic and Upper Jurassic; length ca. 3 m). (After Palmer 1999)

was originally classified with the ornithosuchians (Sect. 8.4.3), which it resembled in possessing a lightly built body and disproportionately large head. Moreover, it was bipedal and probably chased small lizards, killing them with its powerful jaws armed with sharp, recurved teeth. The structure of its skull, cervical vertebrae, and limb joints prove unmistakeably that it was a crocodilian.

Surprisingly, therefore, the ancestors of true crocodiles must have been small, bipedal animals such as *Terrestrisuchus* (Fig. 31b), *Saltoposuchus* (Fig. 31c), and other Saltoposuchidae known from the Upper Triassic rocks of Europe. All these probably fed on small reptiles, insects and other arthropods. Despite their general appearance and insectivorous mode of life, the Saltoposuchidae, like the Sphenosuchidae, possessed a number of diagnostic crocodilian features. These included the elongation of the main wrist bones into rod-shaped elements and a long backwardly pointing spine on the lower part of the pectoral girdle, while the pelvis had an open acetabulum (hip socket). In addition, there were a number of crocodilian specialisations in the skull (Parrish 1997; Benton 2004). *Terrestrisuchus* was smaller and more delicately built than *Gracilisuchus.* Its body was shorter, its limbs long and slim, and its tail almost twice the length of its head and body combined. It roamed over the semi-arid plains of Europe during the Late Triassic, snapping up insects and small lizards with its elongated jaws. No doubt it moved mainly on four limbs but, like its ecological equivalents of today, the frilled lizard (*Chlamydosaurus kingii*; Agamidae), the basilisk (*Basiliscus basiliscus*; Iguanidae) and the collared lizard (*Crotaphytus collaris*; Iguanidae), it would have reached its greatest speed bipedally. It must have sprinted on two legs, not only when in pursuit of prey but – even more importantly – when escaping from predators.

One of the first true crocodiles was *Orthosuchus* from the Lower Jurassic of South Africa. It was a small, lightly built, quadrupedal animal whose hind limbs were longer than the forelimbs and thereby indicated its bipedal ancestry. In Fig. 32, three of these animals are shown feeding at a pool containing fish stranded when floodwaters receded. In 'protosuchians' such as this, the skull was ornamented with irregular pits in the surface of the bone as in modern crocodiles. Several other crocodilian characters in the skeleton have been outlined by Benton (2004). Only four families of Mesozoic crocodiles were fully aquatic: *Teleosaurus* (Teleosauridae; Fig. 31a) and *Metriorhynchus* (Metriorhynchidae; Fig. 30e) are among the better known examples. *Teleosaurus,* from the Lower Jurassic of Europe, was similar in appearance to the gharial (see Sect. 4.6.1). Its powerful jaws were narrow and elongated, the sharp teeth interlocking when the mouth was closed. Its back was heavily armoured, its forelimbs short. *Metriorhynchus* of the Middle Jurassic and its relations, the geosaurs, on the other hand, had dispensed with thick armour and their limbs were transformed into paddles, the hind pair longer than the forelimbs. Their tails had evolved fish-like fins used in swimming. These fins were supported by the tip of the backbone, which tilted sharply downwards – a similar adaptation was independently evolved in the Ichthyosauria (Sect. 5.2.2). Fossils of stomach contents show that *Metrio-*

Fig. 32. *Orthosuchus* (Crocodylia; Lower Jurassic; length ca. 1 m). (After Špinar 1995)

rhynchus fed on ammonites, belemnites, large fish, and pterosaurs. *Geosaurus* was about the same size as *Metriorhynchus,* but even more streamlined. One fossil specimen from southern Germany was outlined by a fine film of carbon which indicates the shape of the fleshy limbs when the animal was alive. It swam by powerful undulations of the body (Massare 1988). It is probable that the metriorhynchids would have had difficulty in walking on land. They may well have hunted cephalopods and fishes by sudden acceleration through the water (Benton 2004).

The Eusuchia ('true crocodilians') appeared in the late Cretaceous, were distinguished by the possession of a secondary palate and were very like modern forms. Only the skull and some skeletal fragments of the most spectacular of these, *Deinosuchus* (Upper Cretaceous), have been found. The skull measured 2 m in length, which suggests an animal 15 m or more in total length if its proportions were the same as those of other crocodiles. It inhabited the swamps of North America where it probably ambushed dinosaurs, especially plant-eating duck-billed hadrosaurs. It had a secondary palate that would have allowed it to open its mouth while submerged without swallowing any water – an advanced feature also found in extant forms. *D. hatcheri* from the Upper Cretaceous of North America probably weighed about 15 tonnes and is the largest of all known crocodiles.

No doubt the four types of locomotion found in extant crocodiles were present in these ancestral forms. There are as follows: (1) sprawling, with the limb joints sticking out sideways; (2) the high walk, which is the normal gait, with the limbs tucked under the body; (3) the belly run, in which the animal reverts to the sprawling posture and slides over the surface of the ground, pushed only by the hind limbs as it escapes rapidly into water; and (4) the gallop, which has only rarely been observed. When small animals are surprised in sleep, they sometimes bound or gallop to the water especially when it is some distance away (Cott 1961). Like plesiosaurs and extant crocodilians, the Mesozoic Crocodylia probably swallowed stones as ballast (Sect. 5.2.1).

The eyes of modern crocodiles and alligators have vertical slit pupils, and the same was probably true of the phytosaurs and early crocodilians. An iris with a vertical pupil aperture is characteristic of primarily nocturnal animals that are also active during the day. When fully closed, a slit pupil is capable of excluding a higher proportion of light than a circular pupil while, when fully open, permitting more light to enter the eye and register on the retina than is possible with a round pupil. A reflective tapetum is also present at the back of the retina. This increases the amount of light that stimulates the visual cells by reflecting the light back again through the retina. For this reason, the eyes of crocodilians appears to glow when illuminated at night when the pupil is dilated. When hunting under water, the eyes of crocodiles are protected by the transparent nictating membranes, but vision may be difficult in dark, muddy water. Crocodiles also have acute senses of smell and hearing. The olfactory areas of the brain are well developed, the inner ears highly developed and the auditory canal extends across the head. This presumably assists in the detection of prey, while sensitivity to a wide range of sound frequencies ensures that the young can be heard calling from within their eggshells at the time of hatching (see Ross 1989).

4.6.3.1
Convergence in Dentition

Where Mesozoic reptiles are represented in the world today by living relatives, as in the case of the turtles and crocodilians, it is often easier to imagine their modes of life than those of taxa with no living representatives. Nevertheless, it is possible to form a shrewd idea of the ways in which extinct animals would have behaved by comparing them with unrelated forms that were probably their ecological equivalents in another era. For instance, most aquatic predatory tetrapods have long, prehensile jaws, armed with pointed teeth that grab the prey and kill it. Few of them have dentitions adapted for cutting the bodies of their prey, as these cannot be used on objects that are floating freely in water. Instead, caniniform or dog-like teeth are employed to hold and kill the prey. In some cases crushing dentition is used to deal with hard-shelled molluscs and crustaceans. Small prey is dismembered by shaking it from side to side: pieces that are thus torn off are swallowed whole. Larger prey cannot be dismembered in this way. Instead, the predator twists pieces off. For example, crocodilians and their ecological equivalents roll over and over in the water until portions of meat have been torn from the carcass of their prey.

Crocodilians are usually assumed to be – and to have always been – carnivorous. They normally possess a uniform dentition of simple, peg-shaped teeth, but a species (possibly of Notosuchida) from the Lower Cretaceous of Malawi

Fig. 33. Undescribed herbivorous crocodilian (Crocodylia; Lower Cretaceous; length ca. 40 cm)

is unique in having a reduced mammal-like dentition of simple incisiform teeth, a canine in the upper jaw, and three complex molariform cheek teeth. In addition, the snout was extremely short compared with that of other crocodilians (Fig. 33). Another very unusual feature was that the articular surfaces of the jaw were elongated, and would have permitted fore-and-aft movement. Internal food processing might well have been related to the possession of a secondary palate which could have allowed breathing to take place during mastication, as in living mammals (Clark et al. 1989). Although these authors realised that the animal must have had a rather specialised diet, they nevertheless considered it to have been an insectivorous predator, perhaps on account of the presence of the incisors and canine teeth. Gillian King (1996), however, suggested that this crocodilian might well have been herbivorous (like many other Mesozoic reptiles). To me, it looks omnivorous. The skull measured about 8 cm in length so the total length of the animal would probably have been at least 40 cm since its snout was so short.

Pincer-like jaws with caniniform dentition have evolved in many different semi-aquatic and aquatic groups including phytosaurs, crocodilians, plesiosaurs, and ichthyosaurs – as well as in fossil birds, modern whales, and so on. Crushing and grinding teeth are characteristic of placodonts, mosasaurs, and walruses. Finally, filter-feeding is found in numerous taxa. This is not surprising because it eliminates at least one trophic level between the predator and the primary producers, and exploits these resources directly and efficiently. It occurred in some chelonians, some plesiosaurs, some pterosaurs, and possibly in mesosaurs (Sect. 4.1) according to Taylor (1987b). Collin and Janis (1997), however, did not consider any of the marine reptiles to have been filter-feeders.

4.6.3.2
Physiology and Behaviour

Extant crocodilians have traditionally been regarded as relatively sluggish animals that spend most of the time lying in wait for prey; but they can, in fact, be extremely active when the situation requires it. They have complex hierarchical systems similar to those found in many species of birds and mammals; male crocodilians participate in ritual territorial displays, and elaborate courtship precedes mating. Both sexes assist the young when necessary in breaking their eggshells, pick up the hatchlings in their mouths and carry them to the water where they are bathed before being allowed to swim ashore, and look after them for a couple of months. 'If crocodiles with their typically reptilian brains, are capable of such complex behavioural patterns, there is no reason to suppose that dinosaurs with similarly developed brains should not have led similarly complex lives' (McGowan 1991). Forms with relatively small brains, such as the sauropods, stegosaurs, and ceratopsians had unhurried and uncomplicated lifestyles.

The unusual trend in the direction of crocodilian evolution appears to have been from land to sea, and thence to fresh water. This must inevitably have involved osmotic problems and it is not known how these were solved except in the Eusuchia. A salt-excreting gland, which was not restricted to the Crocodylidae, was probably already present at an early stage of archosaurian evolution and contributed to dispersal across marine barriers of several taxa whose origin was not marine.

Heat loss is an important factor to marine reptiles, and living Crocodylidae inhabit only warm waters. The same was probably true of their Jurassic and Cretaceous predecessors because there is no evidence that they ever evolved endothermy. This helps to explain the distribution of their fossils (Hua and Buffetaut 1997).

The evolution and distribution of the Mesozoic crocodilians have been reviewed, among others, by Neill (1971), Buffetaut (1989), and Alderton (1991).

5 Plesiosaurs and Ichthyosaurs

5.1 Introduction

The nothosaurs of the Upper Triassic period (Sect. 4.4) very probably gave rise to the plesiosaurs ('near lizards'). The close relationship between the two taxa is revealed by *Pistosaurus* (Fig. 24), the only known member of the nothosaur family Pistosauridae. The head of this marine reptile was similar to that of the plesiosaur, but its palate was typically that of a nothosaur. Fossils of *Pistosaurus* have been found in the Middle Triassic deposits of France and Germany along with those of other Sauropterygia. The Pistosauridae ranged between 20 cm and 1 m in length. They evidently swam using their supple bodies and wide sweeps of their deep tails to produce thrust (Ellis 2003; Benton 2004). All these giant reptiles lived in the sea and not in fresh water. This is known because their fossils have been found in association with those of sea urchins, squids, and other marine animals.

The most serious gap in the fossil record of Mesozoic aquatic reptiles occurs between the Upper Permian and Lower Triassic periods. That is where two major groups – the ichthyosaurs and the sauropterygians – evolved, along with other minor and disparate taxa including the thalattosaurs and askeptosaurs (Carroll 1997). The placodonts (Sect. 4.4.1), nothosaurs (Sect. 4.4.2), and plesiosaurs were all derived from sauropterygian ancestry. The Sauropterygia was a monophyletic clade of diapsid reptiles that invaded the oceans of the world in the Early Triassic period. They first radiated during the Middle Triassic and became extinct during the Late Cretaceous.

In contrast, the ichthyosaurs ('fish lizards') – a distinctive and highly specialised group – lie on the lepidosauromorph line of neodiapsids: they were not sauropterygians. The ichthyosaurs were the most obviously aquatic of all marine reptiles (Bauplan 1: Fig. 17), and no transitional forms between them and their distant terrestrial ancestors have yet been found. Ichthyosaurs arose in the Lower Triassic, radiated in the Middle and Upper Triassic, but were outlasted by the plesiosaurs, dying out somewhat before them in the Upper Cretaceous. They, too, however, were in decline before the Cretaceous extinctions (Sect. 5.5).

5.2 Swimming

5.2.1 Plesiosaurs

The locomotion of the plesiosaurs has formed the subject of at least three different hypotheses, outlined by Taylor (1986), Halstead (1989), McGowan (1991), Ellis (2003) and Benton (2004). Because their tails were relatively small, their paddles large and powerful, it can reasonably be assumed that the latter were used for creating thrust. At first, it was thought that plesiosaurs swam by beating their limbs forward and backward as though they were oars. The paddles would, of course, have had to be 'feathered' or rotated to a horizontal position like the feet of a swimming duck during the backstrokes, since they could not be lifted above the surface of the water like the oars of a rowing boat or skiff. However, the fossils do not indicate that such rotation actually took place.

A solution to the problem was proposed by Robinson (1975) who suggested that the plesiosaurs 'flew' through the water – as do turtles and penguins today. In their case, the paddle is flat, having an aerofoil cross section, like that of a bird's wing, with a rounded leading edge and a tapering rear end. It is tipped at an angle to the horizontal so that, in the downstroke, it generates lift and forward thrust. In the upstroke, the angle to the horizon is tipped in the reverse direction so that a smaller forward thrust and lift force are generated. The tip of the paddle describes a figure-of-eight pattern, and each stage of the cycle produces lift and forward movement (Bauplan III: Fig. 17). More recently, Godfrey (1984) has refined this concept by suggesting that the tip of the plesiosaur paddle may have described a crescent-shaped path (Fig. 34) as do the flippers of sea lions. He claimed that the plesiosaurs would not have been able to have moved their paddles up and down in a figure-of-eight. This was because the pectoral and pelvic girdles were heavy, flattened units of bone connected by a dense series of gastralia or belly ribs to form massive, immovable ventral plates. To these were attached the powerful muscles that operated the limb. At the same time, the limb girdles were too weak for strong vertical movements of the flippers to have been made.

Beverly Halstead (1989) suggested that plesiosaurs could have used all four fins in propulsion because their trajectory through the water was undulating. Consequently. the motion of the rear flippers would not have been affected by the turbulence of the vortices created by the power strokes of the front flippers.

The long necks of many plesiosaurs would have made it difficult for them to swim under water and maintain a straight course, because if they veered slightly to one side the water, in striking their necks obliquely, would have caused them to veer even more. For this reason, they are usually depicted as swimming on the surface. This involves the use of more energy than does swimming under water, but the difference would not have been so great when the plesiosaurs swam slowly. If they fed on worms or clams, using their long

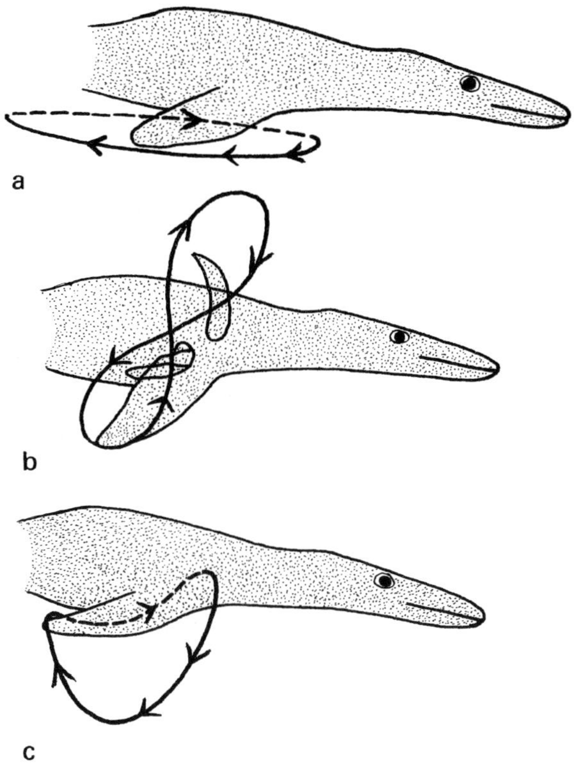

Fig. 34a–c. Plesiosaur locomotion, three hypotheses: **a** rowing under water like a duck; **b** flying under water like a penguin; **c** intermediate style swimming like a sea lion. (Redrawn from Benton 2004 after Taylor 1986). Further explanation in the text

necks to reach down to the bottom, this would have solved the problem. Their piscivorous teeth, however, would have been more suitable for catching fishes and squid-like molluscs. However, these animals were too speedy to have been caught easily – unless the plesiosaurs darted out their long necks to snatch such animals that swam past them (Alexander 1989).

From the fossil records it can be seen that the plesiosaurs, like crocodiles (Sect. 4.6.3), swallowed stomach stones. No doubt these served to counteract the natural buoyancy of the animals. As they were inshore reptiles, finding suitable stones would have caused no problems. Furthermore, the gastroliths may have served to grind up shellfish and similar food items, just as does the grit in the gizzards of grain-eating birds today (Taylor 1992; Ellis 2003). In contrast, ichthyosaurs, which tended to inhabit deeper water far from the shore, did not make use of gastroliths. Not only would it have been more difficult for them to find suitable stones to swallow, but a diet composed largely of squids and fishes would have caused them no digestive difficulties. As we shall see later, many herbivorous dinosaurs did use stomach stones to digest their food (Sect. 10.2).

The discovery of numerous baby and juvenile fossil plesiosaurs in South Australia suggests that this was perhaps a breeding area to which the adults migrated seasonally, like the migration of whales up and down the coast today (Ben Kear in Ellis 2003).

5.2.2 Ichthyosaurs

There can be little doubt that the ichthyosaurs would have been very efficient swimmers (Massare 1988). Their adaptations to marine life were even more extreme than those of the plesiosaurs and their appearance closely resembled that of fishes (Bauplan 1; Fig. 17). They swam as fishes do (Sect. 4.2), their bodies and tails were specialised for propulsion, their limbs were flattened into paddles for balancing, while a fleshy dorsal fin kept their bodies from rolling. Although ecological analogues of porpoises and dolphins, they moved their tails from side to side, not up and down like the flukes of a cetacean.

Christopher McGowan (1991) discussed the mechanics of swimming with special reference to ichthyosaurs. He pointed out that he had initially taken a traditional view of the reversed heterocercal tail as a device for generating a downthrust to overcome the natural buoyancy of an air-breathing vertebrate. (A heterocercal tail is one in which the two lobes are not alike.) When the tip of the vertebral column extends into the dorsal lobe, as in sharks, this is usually larger than the ventral lobe. However, because the lower lobe is deflected more than the rigid upper lobe, its vertical force is greater and the tail generates a net upward thrust. Elasmobranch fishes do not have swim bladders and are therefore heavier than seawater. When not actually moving, they rest on the bottom but, when swimming, the comparatively rigid pectoral fins – whose cross section resembles that of an aeroplane wing – gives lift to the forepart of the body while the heterocercal caudal fin lifts the tail. In contrast, a reversed heterocercal tail would drive the rear part of the body down into the water (Fig. 35).

McGowan (1991), however, wrote that he had changed his mind when he read a paper by Taylor (1987a) who, like Wall (1983), argued that ichthyosaurs may not necessarily have been much lighter than seawater. Many modern aquatic animals, such as otters, hippos and some seals, have densities as great or even greater than that of water. Their weight is incurred by possessing very dense bones. The advantage of being heavy is that these animals do not need to expend energy in order to remain submerged, nor do they need to swallow gastroliths. Although some ichthyosaurs may have had dense bones, the Lower Jurassic forms examined by McGowan did not. In this, they were similar to whales and other cetaceans that must exhale before diving. Taylor (1987a) stressed that the increasing water pressure as an ichthyosaur dived would soon have neutralised any buoyancy. A tail that generated a downwards thrust would thus be disadvantageous. An ichthyosaur swimming at the surface may have initiated a dive by a downward flexure of the tail. This would have produced a downward movement of the head, driving the animal deep into the wa-

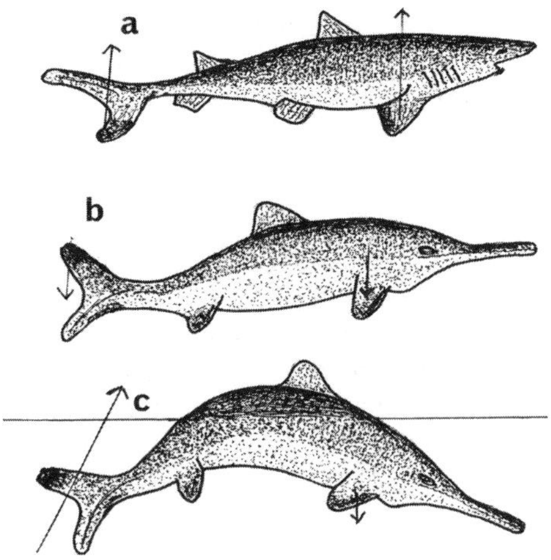

Fig. 35a–c. Ichthyosaur locomotion showing comparison with that of a shark. **a** Shark with excess weight counteracted by caudal and pectoral fins; **b** ichthyosaur swimming horizontally with natural buoyancy counteracted by downward thrust from caudal and pectoral fins; **c** ichthyosaur diving. *Arrows* indicate vertical thrust. Further explanation in the text

ter where changes in the horizontal swimming level would have been effected by altering the inclination of the pectoral fins. The pelvic fins would probably have acted as stabilisers, maintaining the body on a straight course, correcting both up and down as well as side-to-side movements.

The effectiveness of an inclined plane depends on the ratio of lift to the drag force. Both the lift force and the drag increase with increased area, so the lift-to-drag ratio cannot be enhanced merely by altering the size of the plane. A long narrow plane, like the wing of an aeroplane, has a much higher lift-to-drag ratio than does a square plane. The relative narrowness of an inclined plane or wing is expressed by the 'aspect ratio' – the ratio of length to width. For example, if a plane were 10 units long and 10 wide, it would have an aspect ratio of 1. On the other hand, a plane 30 units long and 5 units wide would have an aspect ratio of 6. Planes with high aspect ratios generate less turbulence at their tips than do planes with low. Furthermore, the lift-to-drag ratio is increased by having a streamlined profile (McGowan 1991).

The movements of animals both in water and in air are strongly influenced by the density of the medium in which they are travelling. Although water is over 800 times denser than air, swimming is, somewhat surprisingly, relatively inexpensive in energy. This is partly because water provides more buoyancy than air does. Swimming animals cannot influence the density of the water that surrounds them, but flying animals can do so by selecting the altitude at which they are cruising. (That is why migrating birds usually fly at relatively high altitudes where the density of the atmosphere is lower.)

Since drag increases with the cross-sectional area of the body as well as with the number and size of external projections such as fins, speedy aquatic animals tend to be elongated and to be able to retract the fins or flatten them against their bodies when swimming fast.

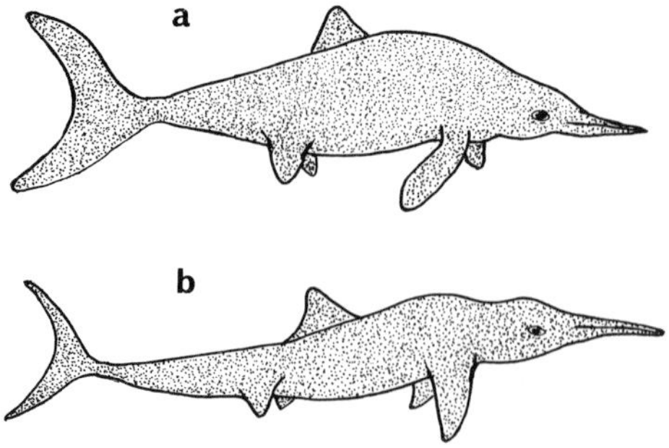

Fig. 36. a *Stenopterygius megacephalus*. **b** *S. hauffianus* (Ichthyosauria; Lower Jurassic and Middle Jurassic; length ca. 3 m). The aspect ratio of length to width was moderately high in *S. megacephalus*, but even higher in *S. huffianus*. Further explanation in text

The pectoral and pelvic fins of the ichthyosaurs were tilted obliquely downwards, like the flight feathers of a dart. They would have acted as inclined planes whenever the body veered from its intended direction, and generated correcting forces to bring it back onto a straight course. Since their orientation was oblique, they would have corrected both vertical and lateral movements (McGowan 1991). A streamlined body with a high aspect ratio was not characteristic of all ichthyosaurs. Some of the early forms, such as *Stenopterygius megacephalus* (Fig. 36a) from the Lower and Middle Jurassic of Europe, had large tails with only moderate aspect ratios. They were clearly not adapted for cruising as were most of the later ichthyosaurs. The congeneric *S. hauffianus* (Fig. 36b), in contrast, was a fast endurance swimmer. Its tail was slender and had an aspect ratio as high as that of a modern swordfish.

Dolphins and porpoises are well known to leap repeatedly out of the water as they swim. This is called 'porpoising' and is necessary for breathing when travelling at high speed. If only enough of the body were exposed to bring the blow hole above the surface of the water, as happens at lower speeds, the turbulence created would be so great that the resulting drag would slow the animals below the critical level for efficient swimming. Moreover, porpoising may actually save energy by reducing drag which is very much greater in water than in air, even at high speed. Although it costs energy to leap, the faster the animal is swimming, the longer the leap and the greater the amount of drag avoided. The calculations necessary to assess the critical speed, above which leaping saves energy, are extremely complicated and the factors involved are not readily estimated. Nevertheless, Alexander (1989) believed that ichthyosaurs porpoised like dolphins. They may also have dived quite deeply to find their prey. Today, the tetrapod that dives most deeply is the sperm whale (*Physeter macrocephalus*), whose deepest recorded dive is 2.25 km (1.4 miles). The stomach contents of another individual included a species of dogfish found only on the seafloor, suggesting that a depth of over 3 km has been reached. The various adaptations for deep diving are both physiological and mechanical; but which, if any of them, evolved among the plesiosaurs and ichthyosaurs is quite unknown.

5.3 Plesiosaurs

The 19th century palaeontologist, W.D. Conybeare, who became Dean of Llandaff, was one of the first to study the plesiosaurs in detail and gave them their name. (He also gave its name to the first of the mosasaurs because it was found near the River Meuse.) Conybeare described plesiosaurs as 'snakes threaded through the bodies of turtles'. This may be an apt portrayal of their general appearance, but it does not help to explain either their relationships or their modes of life. Brown (1981) reviewed their phylogeny and classification.

5.3.1 Plesiosaur Diversification

The first true plesiosaurs of the Upper Triassic period were larger than the nothosaurs to which they were so closely related, if they were not actually descended from them (Sect. 5.1). Their body lengths ranged about 2–15 m. The order Plesiosauria contained four main families: Plesiosauridae, Cryptoclididiae, Elasmosauridae and Pliosauridae (Brown 1981; Ellis 2003; Benton 2004). Some had long necks, others short. The plesiosaurids, known mainly from the Lower Jurassic of Europe, had small skulls and fairly long necks. *Plesiosaurus* (Fig. 37a) is a typical example. The cryptoclidids, such as *Crypto-*

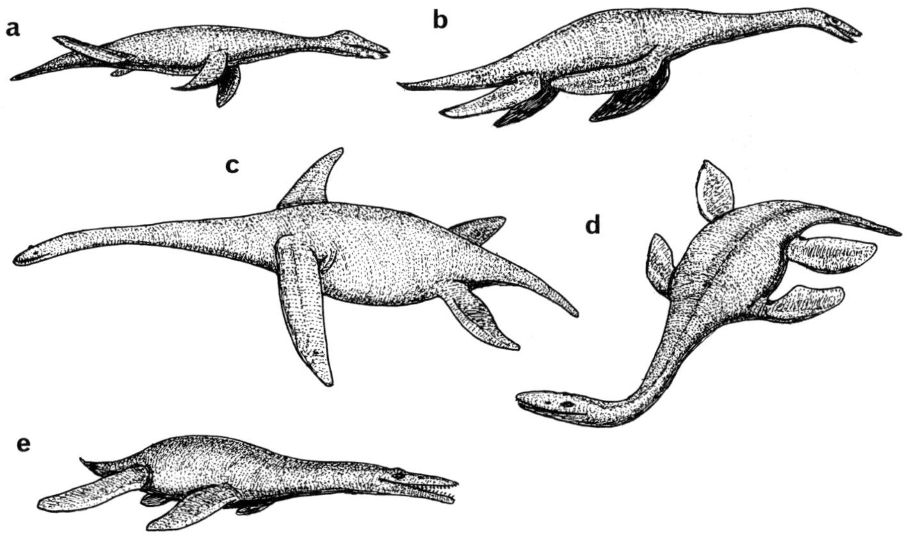

Fig. 37. **a** *Plesiosaurus* (Plesiosauridae; Lower Jurassic; length ca. 2.3 m), **b** *Cryptoclidus* (Cryptoclididae; Upper Jurassic; length ca. 4 m), **c** *Muraenosaurus* (Elasmosauridae; Upper Jurassic; length ca. 6 m), **d** *Macroplata* (Pliosauridae; Lower Jurassic; length ca. 4.5 m) **e** *Peloneustes* (Pliosauridae; Upper Jurassic; length ca. 3 m). (After Palmer 1999)

clidus (Fig. 37b) from the Upper Jurassic of Europe, had even longer necks in proportion to body. There were 30 cervical vertebrae and the skulls had pointed snouts with nostrils set back from their tips. The elasmosaurids from the Upper Jurassic of Europe had longer necks still. For example, in *Muraenosaurus* (Fig. 37c), the neck was as long as the body and tail combined, and contained 44 vertebrae; and *Elasmosaurus* (Fig. 38) from the Upper Cretaceous of Asia and North America had an even longer neck that consisted of 71 vertebrae. These were undoubtedly ambush, or 'sit-and-wait' predators.

In contrast, the pliosaurs had necks that became progressively shorter while the paddles grew larger and more powerful during the course of evolution. The neck of the Lower Jurassic *Macroplata* (Fig. 37d) from Europe contained 29 slightly shortened vertebrae. The slender, crocodile-like skull was only slightly larger in proportion to those of the plesiosaurids. Although smaller than *Macroplata*, the pliosaur *Peloneustes* (Fig. 37e) from the Upper Jurassic of Europe and America, had an even shorter neck with only about 20 vertebrae, and a larger head. It was more streamlined than *Macroplata*, which must have fed

Fig. 38. *Elasmosaurus* (Elasmosauridae; Upper Cretaceous; length ca. 14 m). (Cloudsley-Thompson 1994).

Fig. 39. *Kronosaurus* (Pliosaudridae; Lower Cretaceous; length ca. 5 m). (Cloudsley-Thompson 1999)

mainly on fishes, and its teeth were fewer and blunter. They were better adapted to catching soft-bodied squids and for crushing hard-shelled ammonites. *Kronosaurus* (Fig. 39) from the Lower Cretaceous of Australia is the largest known pliosaur. (The name comes from the mythological Kronos who devoured his children.) Its skull measured about a quarter of its total body length and was larger and even more powerful than that of *Tyrannosaurus rex*!

The plesiosaurids, cryptoclidids, and elasmosaurids fed mainly on molluscs and fishes. As we have seen (Sect. 5.2.1), they would have used their long necks to shoot out their heads after fast-moving prey that swam near them unaware of imminent ambush. Their long, pointed conical teeth interlocked when the jaws were shut – an adaptation for retaining slippery fishes and cephalopods in their mouths when the jaws closed (Benton 2004). In contrast, the pliosaurs were adapted for long-distance cruising and fast swimming. *Kronosaurus* (Fig. 39) and *Liopleurodon* spp. (Fig. 40) from the Upper Jurassic of Europe, with heavy heads, short necks, and streamlined bodies, were highly manoeuverable and certainly not sit-and-wait predators like the earlier pliosaurids and all other plesiosaurs. They probably fed on cephalopods, sharks and other fishes as well as on ichthyosaurs and smaller plesiosaurs. With their powerful jaws and sharp, broad-based teeth, the Pliosauridae had skulls designed like a lattice reinforced with girdles that resisted the great bending moments generated during biting (Taylor 1992). The mandibles were also like box girders and for the same reason, but retained a streamlined shape.

Plesiosaurs almost certainly came ashore in large numbers to lay their eggs in nests excavated in the sand – as do marine turtles today. As already mentioned, the discovery of numerous baby and juvenile fossil plesiosaurs in South Australia suggests that this was perhaps a breeding area to which the adults migrated seasonally (Sect. 5.2.1). When on land, they would have been much more vulnerable to predation from other Mesozoic reptiles than when they were swimming in the ocean; while the newly-hatched young would have faced a perilous journey down the beach to the sea. No doubt safety depended on

Fig. 40. *Liopleurodon* (Pliosauridae; Upper Jurassic; length ca. 12 m; after Palmer 1999) with ichthyosaurs

numbers, so hatching must have been synchronised. In addition to terrestrial reptiles, the enemies of baby plesiosaurs probably included crocodilians and pterosaurs. Adults in the water must have faced mosasaurs (Sect. 4.5), crocodilians, larger plesiosaurs and ichthyosaurs.

5.4 Ichthyosaurs

Ichthyosaurs first appeared in the Lower Triassic alongside the placodonts and nothosaurs. They were the most highly specialised of all aquatic reptiles, and evolved to become top marine predators, yet maintained essentially the same body shape throughout most of the Mesozoic Era – about 150 million years.

Like nothosaurs, thalattosaurs, crocodilians and other amphibious reptiles, plesiosaurs would almost certainly have been ectothermal as are modern crocodiles. These usually sun themselves during the day, taking to the water whenever the body temperature rises to its optimum, and again during the night when the temperature of the air falls below that of the water (Cott 1961; Cloudsley-Thompson 1999). In contrast, the ichthyosaurs may well have been endothermal as some of the larger theropods and dinosaurs probably were (Sect 7.3). Undoubtedly they were extremely active and, if not actually homeothermic, would have generated a considerable amount of endothermal heat when chasing their speedy prey. According to Richard Cowen (in Ellis 2003), they may have breathed in mid-air whilst leaping out of the water – an energetically economical form of high speed locomotion for an aquatic animal (Sect. 5.2.3). Furthermore, they were undoubtedly gregarious, if not fully social, and swam together in large schools as porpoises and dolphins do.

5.4.1 Ichthyosaur Diversification

Of all reptiles, the ichthyosaurs were the most highly specialised for aquatic life. So fully adapted to a marine existence were they that they never came ashore, but gave birth to ovoviviparous young in the sea. The babies emerged tail first, as do those of whales. (If they were born head first, they would drown.) The external nares or nostrils were set back near the eyes so that the animals had only to break the water surface very slightly in order to breathe.

Mixosaurus (Fig. 41; 'mingling lizard'; Mixosauridae) from the Lower and Middle Jurassic strata of Europe, Asia, and North America, was one of the earliest of the ichthyosaurs. It was small, lacked a tail bend, and had primitive limbs with five digits. These were, however, elongated by the addition of many smaller bones (hyperphalangy). The huge *Cymbospondylus* (Fig. 42), also from northern Europe and America, belonged to a different family (Shastasauridae). It was one of the least fish-like of the ichthyosaurs, yet it may have had a slight tail bend and possibly a dorsal fin. On the other hand, *Mixosaurus* had an

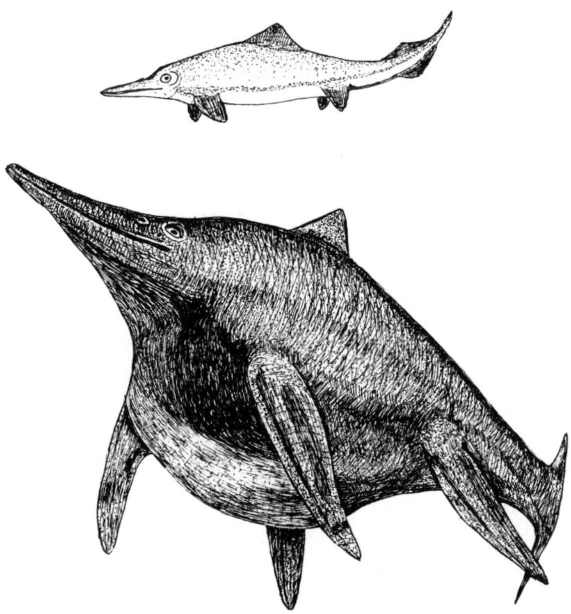

Fig. 41. *Above Mixosaurus* (Mixosauridae; Lower and Middle Triassic; length ca. 1 m). *Below Shonisaurus* (Shastasauridae; Upper Triassic; length ca. 15 m). (After Palmer 1999)

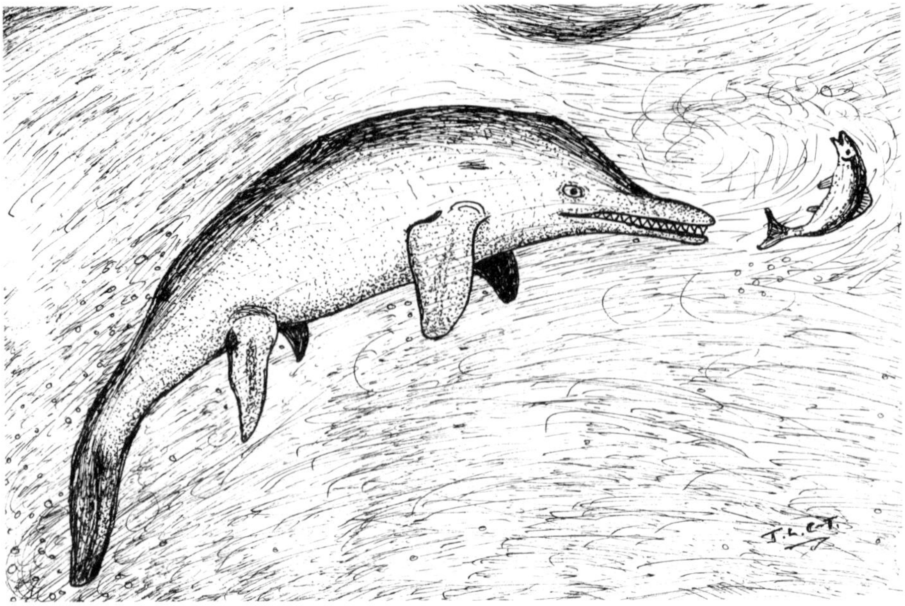

Fig. 42. *Cymbospondylus* (Shastasauridae; Middle Triassic; length ca. 10 m)

unmistakeable dorsal fin and probably the beginnings of a lobe on the upper surface of its tail. The front paddles were longer than those of the hind pair, and the narrow pointed jaws were well armed with numerous sharp teeth well adapted for catching fishes. These teeth were set in individual sockets, as in *Cymbospondylus* and other Triassic ichthyosaurs, and not in a groove. The eyes of *Mixosaurus* were very large, as in later forms (McGowan 1991). The ichthyosaur eye was surrounded by a sclerotic ring of bony plates. These not only helped in focusing, but protected the soft tissues from the pressure of the water when the animal dived deeply. Apart from details of the skull and the fact that the tail fin was not fully developed for swimming, none of these early ichthyosaurs were very much different from their descendants in the Jurassic and Cretaceous Eras.

The massive *Shonisaurus* (Fig. 41; Shastasauridae) from Upper Triassic rocks in Nevada, North America, was the largest ichthyosaur so far discovered. It enjoyed a number of evolutionary advances over *Cymbospondylus*. Its shape was decidedly more fish-like, its backbone bent sharply downwards into the ventral lobe of the tail, and its huge frame was divided into approximately equal thirds, viz head and neck, body, and tail. It also possessed a number of structural peculiarities that separate it from the main line of evolution. Its elongated jaws only had teeth at the front, while its limbs were exceptionally long and narrow. Moreover, they were all of equal size – in most of the ichthyosaurs, the front paddles were longer than the hind pair (Palmer 1999).

Fig. 43. *Ichthyosaurus communis* (Ichthyosauridae; Lower Jurassic; length ca. 2 m). (Cloudsley-Thompson 1994).

The Triassic ichthyosaur genera *Mixosaurus* (Fig. 41), *Cymbospondylus* (Fig. 42), and *Shonisaurus* (Fig. 41) are represented by numerous fossils and are therefore comparatively well known. Less well represented are *Shastasaurus, Californosaurus, Merriamia*, and *Toretocnemus*, all from the Upper Triassic of California and, in the case of *Shastasaurus,* also from Mexico and the Palaearctic region. *Shastasaurus* was the largest of them all and may have exceeded *Cymbospondylus* in length. *Californosaurus* reached 3 m; the other two genera were both smaller (±1 m) – about the same size as *Mixosaurus.* Other Triassic ichthyosaur genera are even less well known. Within each genus, however, some features are primitive, others more advanced. Although Triassic ichthyosaurs were so diverse, two basic types can be distinguished – those with narrow fins, such as *Cymbospondylus, Shonisaurus* and *Toretocnemus,* and those with broad fins, including *Mixospondylus* (McGowan 1991). Similar types were present throughout the succeeding ages.

The numbers of genera in the Jurassic and Cretaceous periods were much larger than in the Triassic. Most of the fossil ichthyosaurs found in England are from the Lower Liassic (the lowest strata of the Lower Jurassic), and a few from the Upper Lias. The former belong to the genera *Ichthyosaurus* (Fig. 43) and *Temnodontosaurus* ('cutting-tooth lizard' = *Leptopterygius*; Fig. 44a), the latter to *Stenopterygius* ('narrow fin'; Fig. 44b) and *Eurhinosaurus* (Fig. 44c). In Germany, on the other hand, the strata richest in ichthyosaurs are Upper Liassic. Several hundred complete fossil skeletons of *Ichthyosaurus* spp. have been found near Holzmaden in the shales of southern Germany. The tiny bones of unborn infants, some actually emerging from the bodies of the females, are preserved in a number of them. In contrast to their Triassic precursors, the Jurassic ichthyosaurs had distinct tails while their numerous pointed teeth were set in a continuous groove and not in individual sockets (McGowan 1991).

Although *Temnodontosaurus* (Fig. 44a) from the Upper Triassic and Lower Jurassic of Europe reached a length of 9 m, most of the Jurassic ichthyosaurs

Fig. 44. a *Temnodontosaurus* (Leptopterygiidae; Upper Triassic and Lower Jurassic; length ca. 9 m), **b** *Stenopterygius* (Stenopterygidae; Lower and Middle Jurassic; length ca. 3 m), **c** *Eurhinosaurus* (Leptopterygiidae; Lower Jurassic; length ca. 2 m), **d** *Ophthalmosaurus* (Ichthyosauridae; Upper Jurassic; length ca. 3.5 m), **e** *Platypterygius americanus* (Ichthyosauridae; Upper Cretaceous; length ca. 7 m). **a, c** After Palmer (1999); **b, d, e** after McGowan (1991)

were smaller. *Temnodontosaurus* was relatively longer and more slender than *Ichthyosaurus.* Its forelimbs and hind limbs were also relatively larger and approximately equal in length, whereas, in *Ichthyosaurus* spp. the forefins were usually up to twice the length of the hind fins. They consisted of at least five, and often six or seven digits. In *Temnodontosaurus,* on the other hand, the fins were relatively narrow and consisted of only three or four digits. This large animal probably cruised in shallow waters, preying upon giant squids and ammonites. In contrast, *Ichthyosaurus* spp. were better adapted to eating fishes. *Ichthyosaurus conybeari* had a pointed rostrum with numerous sharp teeth and large eyes. *I. brevicaps* had a short rostrum so that its large orbits give the skull a remarkably bird-like appearance. Adult *Stenopterygius quadriscissus* (Fig. 44b) was unusual in having very small teeth, although they were well developed in juveniles, while its fin was exceptionally large. In *Eurhinosaurus* (Fig. 44c) and *Excalibosaurus,* the lower jaw was only half the length of the skull. In both these genera, the teeth stuck out sideways along the length of the blade-like upper jaw, as in the modern sawfish (*Pristis zijsron*) and allied species. The jaw may have served to probe in mud and sand and flush out molluscs and flatfish. Finally, *Ophthalmosaurus* (Fig. 44d) was remarkable for its huge eyes whose diameter sometimes reached 120 cm. They occupied the entire depth of the skull on both sides and suggest that their possessor may have fed at night upon squid, the latter preying on plankton-eating fishes (McGowan 1991; Palmer 1999; Ellis 2003).

The fossil record gives a distinct impression that ichthyosaurs were very abundant during the late Middle and Upper Jurassic periods, but lacked diversity. If, as seems likely, this is true and not merely the result of chance fossilisation, it would imply that conditions were remarkably uniform throughout the oceans during that immense period of time. Ecological niches must have been relatively few but extremely spacious.

These conditions persisted well into the Cretaceous period. Most Cretaceous ichthyosaurs appear to belong to a single genus, *Platypterygius* ('broad fin') (Fig. 44e) which was widely distributed. Only about five distinctly recognisable species have been described (McGowan 1972), and even these are based on incomplete material. The species known best are *P. americanus* (Fig. 44e) from North America, and *P. australis* from Australia. These were the last remaining ichthyosaurs before the group became extinct during the latter part of the Cretaceous, about 25 million years before the end of the period.

Platypterygius spp. had small tails in comparison with those of Jurassic forms, and unusually long, narrow fins. They may, therefore, have been adapted for cruising over long distances rather than for sprinting. The fins were robust, and articulated with a large pectoral girdle, so they may have been used for paddling rather than steering (McGowan 1991).

5.5 Conclusions

In general, predatory animals are of two kinds: ambush or sit-and-wait predators, many of which are capable of a sudden dart after their prey, and pursuit predators. Both types occurred among marine Mesozoic reptiles. Ambush predators dominated the large predator communities of the Middle Triassic and Upper Cretaceous periods, whilst pursuit predators were dominant in the Lower Jurassic, as they are among the fishes and whales of today's oceans. Between Late Middle and Upper Jurassic times there was an almost even mixture of ambush and pursuit marine predators, filling nearly all of the feeding guilds (Massare 1997). (A guild is a group of species, all the members of which exploit similar resources in a similar manner.)

The largest Mesozoic marine reptiles were the pliosaur *Kronosaurus* (Fig. 39) and the ichthyosaur *Shonisaurus* (Fig. 41), both of which reached lengths of about 15 m, as we have seen. Now, this is a similar size to that reached by the largest living toothed whale, the sperm whale (*Physeter macrocephalus*) – a fact not without significance. Sperm whales are less than half the length of the blue whale (*Balaenoptera musculus*) which may grow to more than 33 m and weigh over 170 tonnes. Such an immense size could only be achieved by a suspension feeder that continuously filtered large amounts of plankton from its environment. As animals become progressively larger, their nutritional requirements increase exponentially. In very large animals they become so great that plankton is the only renewable source of nutrition that does not become exhausted.

It is thus highly probable that none of the plesiosaurs or ichthyosaurs were filter feeders. This conclusion was also reached by Collin and Janis (1997) who argued that the primitive tetrapod character of an undivided oropharyngeal cavity, the chance of muscular lips and cheeks, a relatively immobile tongue, and the lack of a coordinated swallowing reflex were not conducive to the evolution of suspension feeding. Nevertheless, although unlikely, a posterior secondary palate could possibly have developed among the crocodilians and become a feature of the Mesozoic marine crocodiles (Sect. 4.6.3.1). Morphological constraints rather than scarcity of suitable planktonic prey would therefore have prevented the evolution of mechanisms for filter-feeding. Without the mammalian ability to form a tight seal at the back of the mouth and to collect and swallow small food particles, Mesozoic marine reptiles would have been limited to feeding on larger individual prey items. For this they were well adapted.

At first sight, it might appear strange that the ichthyosaurs, which were so highly adapted to marine life, should have died out before the plesiosaurs. Throughout their existence, however, the ichthyosaurs occupied a more homogeneous environment than that of the plesiosaurs and, consequently, show much less diversity in the fossil record. There are about 38 extant species of porpoise (Phocaenidae) and dolphins (Delphinidae) – not counting the four river dolphins (Platanistidae) – and 31 species of seals, walrus and sea lions that comprise the ecological equivalents of the plesiosaurs. These differences are probably without much significance, but there is undoubtedly more morphological diversity among the Pinnipedia than among the smaller toothed whales (Odontoceti), the ecological equivalents of most of the ichthyosaurs.

The last ichthyosaurs disappeared about 25 million years before the end of the Cretaceous, while plesiosaurs have been found in the late Maastrichtian at the very end of the period, although they were much reduced in numbers and diversity by then. This suggests that gradual climatic change was taking place towards the end of the Cretaceous period. Sea levels apparently dropped comparatively rapidly and the climate became more variable. Pterosaurs and dinosaurs were also becoming less common at the time. Food chains would have been adversely affected throughout the world, and this may well have played a part in the extinction of the giant marine reptiles of the Mesozoic Era (Sect. 12.4).

6 Aerial Mesozoic Reptiles

6.1 Introduction

During much of the Mesozoic Era, the air was dominated by prehistoric reptiles of the order Pterosauria. These archosaurs (Table 2) appeared during the Upper Triassic period, about 228 mya. This was some 70 million years before the first known birds (*Archaeopteryx* spp.). The pterosaurs flourished until the end of the Cretaceous, 155 million years later. Two suborders of pterosaurs are recognised: Rhamphorhynchoidea which had long, stiffened tails, and Pterodactyloidea with short tails. The rhamphorhynchoids evolved first but died out during the Upper Jurassic, somewhat after the pterodactyloids came onto the scene. Fossil pterosaurs have been found throughout most of the world, mainly concentrated in marine deposits.

Peter Wellnhofer (1991) and Norman and Wellnhofer (2000) have not only outlined the probable lines of descent of the pterosaurs, but have also provided illustrations indicating the way in which they may have evolved their wings and powers of flight. Benton (1985) regarded them as archosaurs and a close sister group to the Dinosauria, as had Padian (1984) and many others. According to Wild (1984), the small size, enlarged hooked claws and elongated fourth finger could have been acquired only during a scansorial stage in their ancestry. In his hypothetical prepterosaur, the body surface tended to become enlarged from a lateral band of skin that extended from the front limb and fourth finger to the flanks of the body and the upper parts of the legs. The formation of a flight membrane was beneficial as it checked the descent of these reptiles if they fell from the trees they were climbing. The ability to glide improved as the fourth digit lengthened, the fifth disappeared and the forequarters and pectoral girdle were strengthened. The sternum, too, became ossified and afforded an area for the attachment of powerful flight muscles. This hypothetical prepterosaur has not been found – it probably never will be, because transitional 'missing links' naturally occurred over relatively brief periods of geological history and were quickly replaced by their better adapted descendants.

The earliest pterosaurs, such as *Eudimorphodon* (Fig. 45) from Europe, showed all the unique features of the taxon: viz. a short body, reduced and fused hip bones, five long toes – of which the fifth was divergent – and a long neck under a large but lightweight head with pointed jaws. The reduction in weight was achieved by the two pairs of temporal openings in the diapsid skull

Fig. 45. *Eudimorphodon* (Rhamphorhynchoidea; Upper Triassic; wingspan ca. 1 m)

(Fig. 2). The membranous wing consisted of flaps of skin attached to the enormously elongated fourth fingers of each hand; the first three fingers were short, with strong claws for grasping rocks and vegetation. The fourth finger was connected to the wrist by an elongated metacarpal bone, and the wing membranes were attached to the body on each side of the thighs. A second membrane ran from the bones of each wrist to the animal's neck (Palmer 1999;

Fig. 46. *Quetzalcoatlus* (Pterodactyloidea; Upper Cretaceous; wingspan ca. 12 m)

Benton 2004). *Eudimorphodon* is well known from fossilised remains preserved in Late Triassic rocks of northern Italy. It was evidently an active flier and flapped its wings as a bird does. Although the head was long compared with that of modern birds, the sternum had evolved into a broad, flattened plate to which the powerful wing muscles were attached. Many of the early pterosaurs of both suborders were fairly small – about the same size as pigeons – but some of the Upper Cretaceous pterodactyloids were enormous. *Quetzalcoatlus* (Fig. 46) from the Late Cretaceous of North America – appropriately in Texas – had an extended wingspan of about 1.2 m. This is comparable with that of an 'executive jet'; and *Quetzalcoatlus* was the largest flying animal ever known (Lawson 1975). Before discussing the pterosaurs, however, brief mention should be made of other aerial Mesozoic diapsid reptiles.

6.2 Gliding Reptiles

Apart from the pterosaurs, no reptiles have evolved flapping flight. Since the Upper Permian, however, small diapsid gliding reptiles have appeared in the fossil record. One of these, *Daedalosaurus* from Madagascar, was so named by Robert Carroll after Daedalus who, in Greek legend, escaped from Crete with the aid of wings which he had made for himself. *Daedalosaurus* is now considered to be a junior synonym of *Coelurosauravus* (Figs. 47, 50a), as is *Weigeltisaurus* from Germany and England.

Fig. 47. *Coelurosauravus* (Squamata; Upper Permian ; length ca. 60 cm). (After Evans 1982)

Fig. 48. *Kuehneosaurus* (Squamata; Upper Triassic; length ca. 75 cm). (Cloudsley-Thompson 1994 after Charig 1979)

The specialised postcranial skeleton of these reptiles had very long ribs that undoubtedly supported a membrane of skin. This would have been opened to enable the animal to glide, and folded back against the body when it was moving through vegetation or on the ground (Evans 1982). Wellnhofer (1991) and Norman and Wellnhofer (2000) described several similar forms, including *Kuehneosaurus* (Figs. 48, 50b) and *Icarosaurus* (Figs. 49, 50c) from the Upper Triassic strata of what are now Britain and USA respectively. The latter was named after Icarus, the son of Daedalus. Like those of *Coelurosauravus*, the membranes with which these animals glided was supported on elongated ribs - of which there were only 10 or 11 pairs, compared with 21 in the case of the

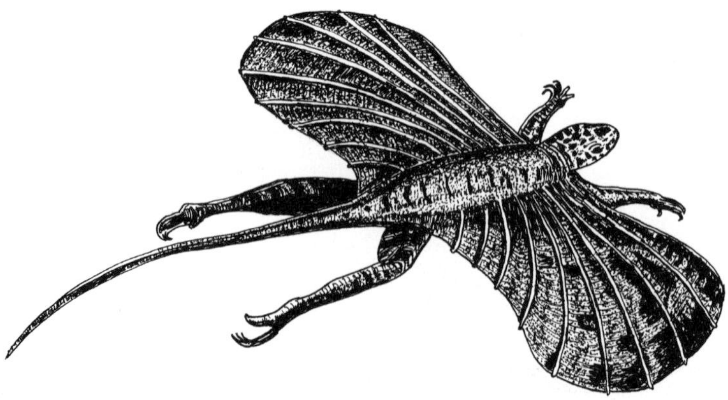

Fig. 49. *Icarosaurus* (Squamata; Upper Triassic; length ca. 30 cm). (After Wellnhofer 1991)

Coelurosauravidae. The wings would, therefore, have been relatively narrower in these Mesozoic forms, than they were in the Upper Permian gliding reptiles.

A similar development of the ribs to support a gliding membrane or patagium is seen in the modern *Draco volans* (Agamidae) and related species. In *Draco,* the five to seven pairs of elongated limbs are usually kept folded, but are spread apart when the animals are gliding. These lizards inhabit the rainforests of Indonesia and feed upon ants and other insects living on the trunks of the tall trees: they eat whilst climbing up one tree and then glide, swiftly and safely, to another. Distances of up to 60 m, have been recorded from a starting height of only 10 m (Evans 1982). Not only does gliding flight save energy, but it enables the lizards to avoid walking across the potentially dangerous floor of the forest. Moreover, the patagium is brightly coloured, and aids camouflage by disrupting the outlines of the creatures. The same was probably true in the case of the Permian and Triassic forms discussed above.

Fossils of an Upper Triassic reptile, *Sharovipteryx mirabilis* (Fig. 50d) (Squamata) show that this, too, was a glider – although somewhat different in form from those already described. It was slender, with a relatively long neck and extremely elongated hind legs and tail. Its hind limbs were more than three times the length of the forelegs. Their proportions exceeded those of modern agamid lizards which run and jump with the trunk elevated so that the front legs may, or may not, touch the ground between strides (Cloudsley-Thompson 1999). This gives a clue as to the probable evolutionary origin of the flight mechanism. The main gliding membrane stretched between the hind limbs and the tail. The long

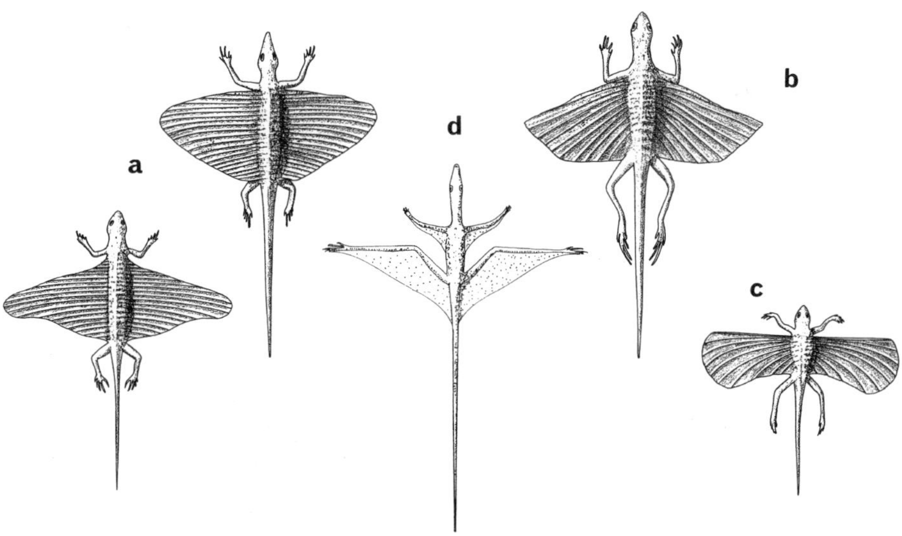

Fig. 50a–d. Gliding lizards (Squamata). **a** *Coelurosauravis* (Upper Permian; length ca. 60 cm), **b** *Kuehneosaurus* (Upper Jurassic; length ca. 75 cm), **c** *Icarosaurus* (Upper Triassic; length ca. 30 cm), **d** *Sharovipteryx* (Lower Triassic; length ca. 25 cm). (After Wellnhofer 1991)

Fig. 51. *Longisquama* (?Pseudosuchia; Upper Triassic; length ca. 15 cm)

tail would have balanced that part of the body extending in front of the membrane. Narrow gliding membranes may well have also been present behind the forelimbs. Although *Sharovipteryx* would not have been able to flap its leg membranes like wings, it could well have glided with the forelimbs stretched slightly forward and the back legs spread out at right angles to the body.

Another curious gliding reptile, described by A.B. Sharov in 1970 from the Late Triassic of Kirghistan, was *Longisquama* (Fig. 51). The enormously elongated, scale-like appendages along its back were actually longer than the body and were arranged in a double row. They could apparently have been held upwards like the wings of a butterfly, or folded down along the sides of the body to form horizontal gliding surfaces. Each individual appendage consisted of a long, thin shaft that broadened towards the distal end and bent backwards. Each pair of these appendages probably corresponded to a dorsal vertebra. The ten pairs of appendages, overlapping at their edges, would have created continuous wing surfaces. *Longisquama* appears to have been a pseudosuchian, and was therefore completely unrelated to the other Mesozoic gliding reptiles discussed above. It acquired its aerial abilities by parallel evolution. The suggestion has been made that the appendages might also have been thermoregulatory devices like those of some pelycosaurs (Sect. 2.5), or display structures that were used to attract mates and threaten rivals. Evans (1982) commented that, if Sharov's description is correct, then any resemblances between *Longisquama* and the Upper Permian flying reptiles must be due to convergence, as indeed is the resemblance between them and *Sharovipteryx*, the other Triassic gliding reptiles, and modern *Draco* spp.

6.3 Pterosaur Flight

During the course of evolution, the oxygen content of the atmosphere has increased and decreased many times. When there was double the amount found today, this might possibly have allowed for the growth of giant plants and reptiles during the Mesozoic Era and also have contributed to the evolution of aerial reptiles. Some of these were very large indeed compared with modern birds.

Three types of flight, which are by no means mutually exclusive, have evolved in the animal kingdom. These are gliding, soaring and flapping flight. For a long while it was assumed that the pterosaurs were merely gliding and soaring animals, and would have been incapable of flapping flight, since their sternum lacked a prominent keel like that of birds. Recent studies, however, have shown that this was not the case. The pterosaurs were almost as well adapted to flapping flight as modern birds are (Wellnhofer 1975, 1991; Padian 1984; Padian and Rayner 1993; see also Benton 2004.)

Pterosaurs had hollow, lightweight bones, streamlined heads, and other aerodynamic adaptations. They may also have possessed a system of air-sacs like that of birds. Moreover, they were probably endothermal, and their bodies covered with insulating hair. Only endotherms have external insulation, and endothermy would have enabled the pterosaurs to maintain the high, sustained metabolic rate necessary for powered flight. Even in the 19th century, it was conjectured by T.H Huxley that the pterosaurs might have been warm-blooded; and as long ago as 1870 H.G. Seeley suggested that they might have had a covering of hair, like that of bats. *Sordes pilosus* (Fig. 52), of which the first specimen was discovered in 1971 among the Late Jurassic deposits of Kazakhstan, seems to have been covered with dense insulating hair except on the tail and legs (de Ricqlès 1975). The exact nature of this hair has yet to be de-

Fig. 52. *Sordes* (Rhamphorhynchoidea; Upper Triassic; length ca. 50 cm). (Cloudsley-Thompson 1978, after de Ricqlès 1975)

termined although, like feathers, it probably evolved from reptilian body scales. Microscope examination of the wing membranes of several species of pterosaurs shows that they were reinforced with stiff fibres, especially distally. These probably prevented the outer parts of the wings from bending. Much of the pelt of *Sordes* has been shown to consist of similar fibres, although true hairs have been detached from areas that were not concerned with flight (Padian and Rayner 1993; Unwin and Bakhurina 1994). Probably all the pterosaurs had hairy coats (see Naish and Martill 2003).

The power stroke in pterosaur flight was directed down and forward, while the recovery stroke was upwards and backwards. When viewed from the side, the path of the wing tip would have described a figure-of-eight. The power stroke was generated by massive pectoral muscles, the upstroke by muscles which ran from the sternum over pulley arrangements at the shoulder joints to the dorsal sides of the humerus bones. Although situated below the wings, these muscles actually pulled them up. A similar system is found in birds.

Pterosaurs had relatively large wings and, in consequence, flew rather slowly. Nevertheless, they were highly manoeuverable: their wings were comparable with those of soaring birds such as vultures and albatrosses, or with aerial predators such as falcons and gulls. They probably took off from trees and cliffs, or jumped into the air after a short run into the wind – as storks and vultures do. Even the larger pterosaurs, such as *Pteranodon* spp. (Fig. 53), could probably become airborne at very low air speeds. Landing must have been

Fig. 53. *Pteranodon* (Pterodactyloidea; Upper Cretaceous; wingspan ca. 7 m). (Cloudsley-Thompson 1994)

awkward for larger species, as it is for large birds, so that the reinforced pelvis and sacrum would, from time to time, have had to withstand quite powerful blows on impact (Benton 2004).

Unwin and Bakhurina (1994) emphasised the bird-like construction of the narrow, stiff wings of *Sordes* which were either free from the legs or else – and most people agree with this – the flight membranes were attached to the thighs, and the legs were intimately involved in the flight apparatus. Unique among flying vertebrates, *Sordes* and other pterosaurs had a structurally non-homogeneous flight surface, with a stiffened outer half and a softer, more extensible, inner region. In *Pterodactylus* and probably other pterodactyloids, a small 'uropatagium' stretched between the inner sides of the legs and the body. The femora were held almost perpendicular to the spinal column – resulting from their association with the uropatagium – and, in consequence, the flight membrane would have been highly manipulatable and the wing loading low. However, the uropatagium must have seriously impeded movement on the ground (Sect. 6.4). At least some fossils reveal the presence of webbing between the three free fingers of the hand supporting a 'propatagium' even larger than is often depicted. Webbing between the toes implies use of the feet as paddles or air-brakes, as reviewed by Naish and Martill (2003).

The suggestion has been made that the cranial crests of *Pteranodon ingens* (Fig. 63g) and *P. sternbergi* (Fig. 63h) might have compensated for air pressure on the beak when the animals turned their heads sideways to the wind. However, not all species of *Pteranodon* had large cranial crests, so clearly these were not essential. Still, they may have contributed to the success of those species that did possess them. Langston (1981) and others have suggested that the function of cranial crests might have been to compensate for the lack of a tail in pterodactyloids. But again, by no means all species had them. Some years earlier, Bramwell and Whitfield (1974) carried out an experimental investigation of the biomechanics of *Pteranodon* (Fig. 53). By means of wind-tunnel experiments on model heads, they showed that the large and long cranial crest was primarily a weight-saving device. By balancing the aerodynamic load on the beak, it allowed the neck muscles to be reduced, thereby saving much more than its own weight. Previously, as already mentioned, the function of the sagittal crest had been thought to counteract the twisting effect of wind pressure on the large, toothless beak. Perhaps both functions were served simultaneously. On the other hand, the crests may have had nothing to do with aeronautics. They could have served in sexual display or for thermoregulation, as with the plates on the backs of stegosaurs (Sect. 7.5.2; Alexander 1989). Even the basal groups of pterosaurs included crested species. Crests were extremely common among the later forms, and were often composed mainly of soft tissues. In *Tapejara* there was a posteriorly directed spike at the back of the skull and a long rod growing up from the tip of the beak. Soft tissue spanned the space between the two spikes, producing a crest larger than the rest of the skull. *Nyctosaurus* had a bizarre rod-like crest even larger than the skull and body combined. Moreover, the rod bifurcated near its base to product a pseudo-antler! (Bennett 2003).

With a sinking speed of 0.42 m s^{-1}, and a flying speed of 8 m s^{-1}, *Pteranodon* would have been able to soar on weak thermals or by hill lift at very light wind speeds. Out to sea, it would have used weak thermals generated by convection; dynamic soaring or slope-soaring over the waves would not have been possible for such a low-speed glider. Its low stalling speed would, however, have enabled it normally to land very gently. The Cretaceous climate was more uniform than that in present times, and light winds prevailed. Although primarily a glider, *Pteranodon* was nevertheless capable of level flight. It could have taken off either by dropping from the cliff on which it roosted or, in emergencies, by facing the wind and just spreading its wings while on a small elevation or on the crest of a wave.

6.4 Locomotion on the Ground

Palaeontologists generally concur that pterosaurs were capable of active flight even though the leverage of their pectoral muscles, in the absence of a keel, was less than it is in birds. But there is somewhat less agreement as to how they managed to walk on land. Indeed, the problem of how they moved on the ground, and took off into the air, dates almost from the discovery of the first fossil remains of the animals. The argument has often been polarised with two contrary views: bipedal versus quadrupedal.

6.4.1 Bipedal Locomotion

Padian (1984), Bennett (1990), and Padian and Rayner (1993) argued that the pterosaurs could have walked well on fully erect hind legs. According to these authors, the pelvic girdle was fixed firmly beneath the body, and the limb movements must have been like those of small bipedal dinosaurs. When running, the wings were held horizontally beside the body (Fig. 54). This view had been proposed as long ago as 1922 by Carl Stieler, with respect to the long-tailed pterosaur *Dorygnathus* from the Liassic of Germany. Stieler concluded that *Dorygnathus* was capable of reaching the speed necessary for take off by running on its toes with short steps and its legs held wide apart. The central issue in the discussion of position has been the anatomy of the pelvis – whether the right and left halves were fused, as they were in dinosaurs or whether they became disjointed vertically. If they were fused, the hip sockets faced outward and lightly downward, then the posture could have been erect (Padian 1984). In *Pteranodon* (Fig. 53), the two halves were indisputably joined, and possibly in *Anhanguera* (Fig. 54) also.

McGowan (1991) considered that too much has been made of pelvic fusion. 'The equation that has been used is: no fusion = hip socket faces outward and upward = laterally splayed hind legs = no bipedal locomotion'. Yet birds lack a fused pelvis and may have hip sockets that face outward and upward 'though

Fig. 54. *Left Rhamphorhynchus* (Rhamphorhynchoidea; Upper Jurassic; wingspan ca. 1 m). *Right Anhanguera* (Pterodactyloidea; Lower Cretaceous; wingspan ca. 4 m). In both cases the reconstructions show possible quadrupedal locomotion above and bipedal locomotion below

admittedly, the head of the femur is somewhat different from that of pterosaurs'. He added that the functional reason why birds lack a vertically fused pelvis has to do with the evolution of balance, and there do not appear to be very good grounds for ruling out the possibility that pterosaurs may have been bipedal. Nevertheless, all the most recent evidence points to quadrupedal locomotion (Sect. 6.4.2).

6.4.2
Quadrupedal Locomotion

Among many others, Wellnhofer (1991) argued that the hind limbs of the pterosaurs pointed sideways in an awkward, sprawling posture. These creatures would, therefore, have scrambled about on all fours (Figs. 54, 55), using their feet and the claws of their hands with the wing tips sticking upwards on either side of the head. This gait has been confirmed independently from several examples of fossilised pterosaur tracks. Some of these are open to question, but most are unequivocal. In one case, an apparent pterosaur track was later proved to have been made by a crocodilian, in another by a horseshoe crab (Xiphosura). Not only do some fossils from Brazil prove that a semi-sprawling posture was adopted, but this has been confirmed by biochemical modelling. Like *Pterodactylus* (Fig. 55) and *Pteranodon* (Fig. 53; Sect. 6.6), the

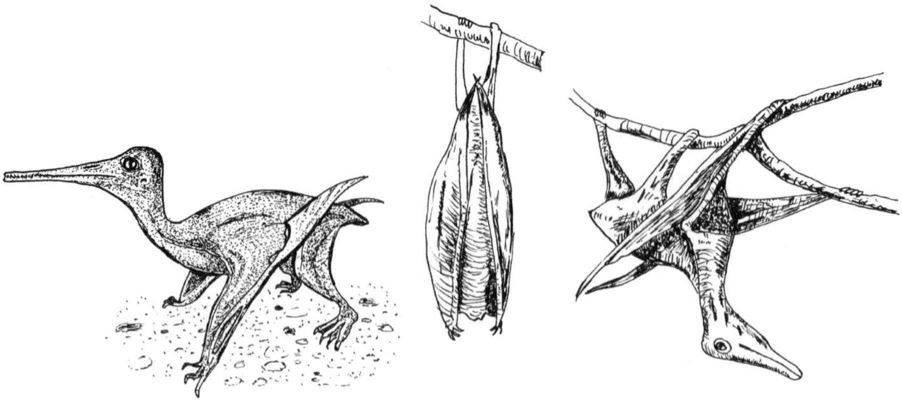

Fig. 55. *Pterodactylus* (Pterodactyloidea; Upper Jurassic; wingspan ca. 75 cm). Reconstructions showing: *left* quadrupedal locomotion. *Centre* Roosting beneath a branch. *Right* Climbing upside down among the branches of a tree. (After Wellnhofer 1991 from Abel 1925)

majority of species probably roosted hanging from the edges of cliffs, or clinging beneath the branches of trees.

Pterosaurs must have spent most of their lives airborne and were presumably rather helpless on the ground. They could probably have taken off from the ground, however, by standing on their hind legs, facing the wind, and stretching out their wings. Some earlier genera such as *Dimorphodon* (Fig. 57) as well as *Pterodactylus* (Fig. 55), which had relatively long hind legs and were small and light in weight, may well have raised themselves into the air by a jump and a simultaneous stroke of the wings (Wellnhofer 1991). The great pterodactyls of the Cretaceous, however, would not have been able to do this. They probably took off from a hanging position on the edges of cliffs and crags, as Bramwell and Whitfield (1974) postulated in the case of *Pteranodon* (Sect. 6.3). Whether they could also have hung downwards from branches and rocky protrusions, as bats do, seems improbable. Certainly the Rhamphorhynchoidea, with their long stiff tails, would have found it very difficult to do so! Many questions about the behaviour of pterosaurs remain to be answered, but it does seem that parallel evolution has occurred between pterosaurs, birds, and bats. The problem is to determine in what ways the pterosaurs most resembled birds, and in what ways bats.

6.5 Rhamphorynchoids

The rhamphorynchoids were the earliest and most primitive of the pterosaurs. Yet these had become already advanced fliers by Upper Triassic times (ca. 228 mya). *Eudimorphodon* (Figs. 45, 56; Sect. 6.1) is a well-known example. Its long tail would have been held out rigidly during flight, counterbalancing the

Fig. 56. *Left Peteinosaurus* (Rhamphorhynchoidea; Upper Triassic; wingspan ca. 60 cm). *Right Eudimorphodon* (Rhamphorhynchoidea; Upper Triassic; wingspan ca. 1 m)

comparatively heavy head and neck. The vertical diamond-shaped flap at the end probably served as a rudder for steering as the animal flew low over the sea searching for the fishes on which it preyed. The jaws were fairly short and the teeth were of two kinds. Those in front were long and peg-like, while the back teeth were short and broad. In *Peteinosaurus* (Fig. 56), which existed roughly contemporaneously but was rather smaller, a large number of small teeth were set behind a few big ones in front. *Peteinosaurus* was even more primitive than *Eudimorphodon* and probably insectivorous. Its wings were short – only twice as long as the hind legs. In other pterosaurs they were nearly three times longer. Its successor, *Dimorphodon* (Fig. 57) from the Lower Jurassic of Europe, had a disproportionately large head, as was typical of the rhamphorhynchoids. The skull was uniquely shaped, like that of a puffin, and was about a quarter as long as the body. The short, sharp teeth may have been an adaptation to a piscivorous diet; while the deep, narrow head could possibly have been used in territorial and courtship display.

Rhamphorhynchus (Fig. 58), from the Late Jurassic of Europe and Africa, has yielded some exceptionally well-preserved fossils of several species in the fine-grained limestone strata of Solnhofen in southern Germany. Microscope study of the wings reveals that thin fibres ran between the fronts and the backs of the wings and strengthened them as do the radiating fingers that support the wings of bats. (These fibres were already mentioned in Sect. 6.3.) The jaws of *Rhamphorhynchus* were long and narrow, and its sharp teeth pointed outwards like the barbs of a fishing spear. Their function is confirmed by the pres-

Fig. 57. *Dimorphodon* (Rhamphorhynchoidea; Lower Jurassic; wingspan ca.1.2 m). (After Augusta 1961)

ence of fish remains in the crop and stomach of some fossil specimens. *Rhamphorhynchus* probably skimmed over the water mopping up fishes, whilst its long tail was held out for stability (Palmer 1999). It would also have poked its elongated snout into the burrows of polychaete worms to extract their occupants (Bakker 1987).

Also from the Late Jurassic of Europe was *Scaphognathus*. Although similar in size to *Rhamphorhynchus*, this genus had a shorter head and long teeth, while the tips of its jaws were blunted. The skull of one specimen has been sufficiently well preserved for the brain cavity to be studied. This was found to be proportionately much larger than that of most other reptiles of similar size – almost as large as that of a modern bird. The cerebellum and associated lobes of the brain were very well developed, which indicates that the animal would have been extremely agile. It apparently had excellent eyesight but a poor sense of smell (Langston 1981). *Scaphognathus* (Fig. 59) was a close relation of *Sordes* (Fig. 52, Sect. 6.3). *Anurognathus* (Fig. 59) differed from other known rhamphorhynchoids in having a relatively short tail. In this it foreshadowed the Pterodactyloidea, although in other respects its body proportions were

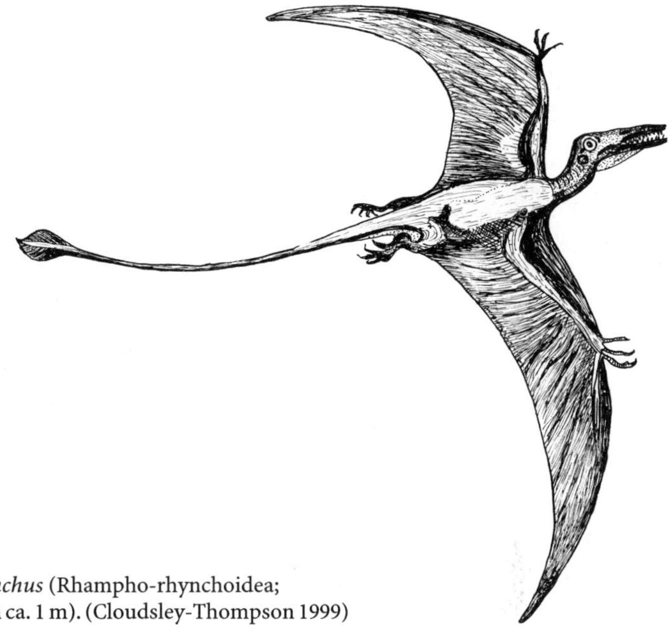

Fig. 58. *Rhamphorhynchus* (Rhampho-rhynchoidea; Upper Jurassic; wingspan ca. 1 m). (Cloudsley-Thompson 1999)

Fig. 59. *Left Anurognathus* (Rhamphorhynchoidea; Upper Jurassic; wingspan ca. 30 cm). *Right Scaphognathus* (Rhamphorhynchoidea; Upper Jurassic; wingspan ca. 1 m). (After Wellnhofer 1991)

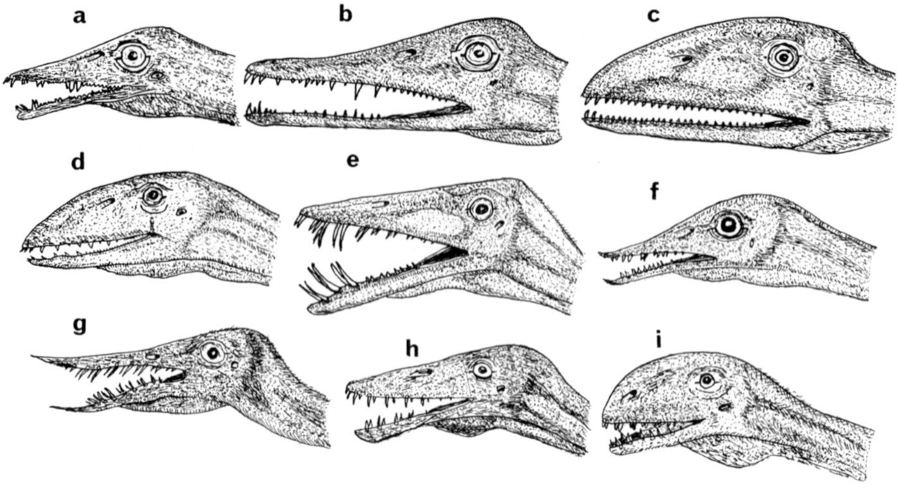

Fig. 60a–i. Heads of rhamphorhynchoid pterosaurs (not to scale). **a** *Eudimorphodon* (Upper Triassic), **b** *Preondactylus* (Upper Triassic), **c** *Peteinosaurus* (Upper Triassic), **d** *Dimorphodon* (Lower Jurassic), **e** *Dorygnathus* (Lower Jurassic), **f** *Campylognathoides* (Lower Jurassic), **g** *Rhamphorhynchus* (Upper Jurassic), **h** *Scaphognathus* (Upper Jurassic), **i** *Anurognathus* (Upper Jurassic). (Mostly after Wellnhofer 1991)

similar to those of the Rhamphorynchoidea; and it was contemporary with *Scaphognathus. Anurognathus* was a very small, slender pterosaur with unusually long wings and a short, deep head. Its teeth were peg-like and it was almost certainly insectivorous. Having dispensed with a long tail, it would have been extremely manoeuverable, snapping up flying insects on the wing.

Much can be learned about the animals that possess them by the study of skulls and teeth. In Fig. 60, reconstructions are shown of the heads of several rhamphorhynchoid pterosaurs. *Eudimorphodon* (Fig. 60a) was a fish-eater as we have seen. *Preondactylus* (Fig. 60b), among the earliest rhamphorhynchoids in the fossil record, may have been ancestral to *Dorygnathus* (Fig. 60e). It is not clear whether it fed on insects or on small fishes. A fossilised gastric pellet of its accumulated bones is interpreted as having been spewed up by a large predatory fish. This could have caught the pterosaur as it was skimming over the surface of the water or, alternatively, the animal might already have drowned.

Peteinosaurus (Fig. 60c) was smaller than *Eudimorphodon*. Its dentition consisted of large anterior fangs followed by a series of smaller tusks. It might have caught insects while on the wing, as bats do. *Dimorphodon* (Fig. 60d) had long, sharp, curving front teeth. These pointed directly upward from the lower jaw and forwards from the upper. The animal's strong bite must have been delivered by a rapid snap of the jaws as it skimmed over the water, snatching fishes from the surface. In addition, however, the neck of *Dimorphodon,* unlike that of many other rhamphorhynchoids, was constructed in such a way that this pterosaur could deliver rapid lunges to grab at prey. Extra large fishes and squids could have been swallowed because, at the midpoint of the lower jaws, there was a zone of weakness. This might have allowed the jaws to bow outwards so that the meal could slip down the throat of the animal (Bakker 1987).

Campylognathoides (Fig. 60f) had the teeth of a piscivore, whilst its unusually large eyes suggest that it might have been adapted for nocturnal hunting. *Rhamphorhynchus* (Fig. 60g) was also a fish-eater, as mentioned earlier. Fossils of several species in this genus, displaying a range of sizes, have been discovered. Fish and squid can detect the dive of a pelican or a puffin just before the bird strikes the water and, consequently, scatter in all directions at the last moment. The hunting methods of the pterosaurs evolved to maximise the rapidity of their strike. Some responded like *Dimorphodon*. In *Rhamphorhynchus*, the S-shape of the neck was accentuated so that the head could be coiled tightly against the shoulders and then shot forward to capture prey. As already mentioned, the jaws and teeth were shaped like fishing spears; even the tips of the jaws were tapered to sharp points which could be used to impale the prey, whose struggles served only to drive the barb-like teeth into its body (Bakker 1987).

Finally, *Scaphognathus* (Fig. 60h) had strong piscivorous fangs, somewhat similar to those of *Dorygnathus* (Fig. 60e), while *Anurognathus* (Fig. 60i) had rounded jaws studded with peg-like teeth. Like *Peteinosaurus*, it must have caught insects on the wing. (The dentition and food of the pterodactyloids is discussed in Sect. 6.6.)

6.6 Pterodactyloids

The first fossils of short-tailed Pterodactyloidea were discovered in the Upper Jurassic strata of Europe. They were assigned to the genus *Pterodactylus* (Fig. 55; 'flight finger'), and possessed all the typical features of pterosaurs. Many different species of *Pterodactylus* have been found, varying mainly in size and the shape of the head. Their narrow jaws and sharp teeth suggest that they must have been mainly piscivorous, like most of the rhamphorhynchoids.

Nothing is known about the direct ancestors of the pterodactyloids, but it can be presumed that they were descended from rhamphorhynchoids and evolved during the Early or Middle Jurassic. They were probably more adroit fliers than their ancestral clade which may eventually have lost out in competition with them. Whereas many of the Rhamphorhynchoidea would mainly have flown with their heads and necks extended, as shown in my earlier drawings, most of the pterodactyloids balanced themselves in flight by curling their long necks backwards, as pelicans do (see below). Another advantage of their behaviour has already been discussed in the case of *Dimorphodon* and *Rhamphorhynchus* (Sect. 6.5), which used their long necks to lunge forward when they capture prey.

Fig. 61. *Left Germanodactylus* (Pterodactyloidea; Upper Jurassic), roosting beneath a branch (wingspan ca. 1.1 m). *Right Gallodactylus* (Pterodactyloidea; Upper Jurassic; wingspan ca. 1.4 m). (After Wellnhofer 1991)

In addition to more than a dozen species of *Pterodactylus*, Upper Jurassic Pterodactyloidea included species of *Germanodactylus, Gallodactylus, Ctenochasma, Gnathosaurus, Huanhepterus, Dermodactylus,* and *Mesadactylus* (Wellnhofer 1991). The pterodactyloids radiated even more extensively during the Cretaceous (see below).

Germanodactylus and *Gallodactylus* were contemporaries in the Upper Jurassic of Europe, although *Germanodactylus* has also been recorded from older strata; but neither of these genera survived into the Cretaceous period. The typical characteristics of *Germanodactylus* spp. (Fig. 61) were low crests of bone along the centre of the skull, beginning above the nostrils and extending over the eye sockets. *Germanodactylus* spp. had long rows of short and powerful teeth. In *G. cristatus,* the tips of the jaws were toothless (Fig. 62a). They were probably covered by pointed, horny beaks. *Gallodactylus* (Fig. 62b), on the other hand, had an elongated jaw with only a few slender teeth at the front of the beak. These pointed forwards and would have grasped slippery fishes most effectively.

Ctenochasma is a rare European genus from the Solnhofen limestone of Bavaria. Its long, slender jaws (Fig. 62a) contained over 250 teeth in the adults. They formed a strainer with which items of food must have been filtered from the sea. Presumably these pterosaurs either swam, or waded in shallow water as

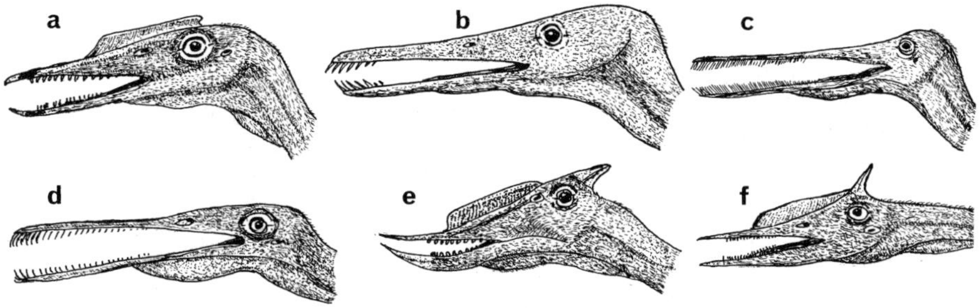

Fig. 62a–f. Heads of Jurassic pterodactyloid pterosaurs (not to scale). **a** *Germanodactylus* (Upper Jurassic), **b** *Gallodactylus* (Upper Jurassic), **c** *Ctenochasma* (Upper Jurassic), **d** *Gnathosaurus* (Upper Jurassic), **e** *Dsungaripterus* (Upper Jurassic and Lower Cretaceous), **f** *Phobereptor* (Upper Jurassic and Lower Cretaceous. (Mostly after Wellnhofer 1991)

many shore birds do today. *Gnathosaurus*, also from the Solnhofen limestone, is likewise extremely uncommon in the fossil record. It, too, has long and slender jaws (Fig. 62d), but there were only about 130 teeth for filtering food. However, *Ctenochasma* was a rather small pterosaur while *Gnathosaurus* had a wingspan of ca. 1.7 m and would have strained much larger food items through its teeth.

Dsungaripterus (Fig. 62e) and *Phobereptor* (Fig. 62f) both arose in the Upper Jurassic and extended into the Cretaceous. Fossils of the former genus have been found in the Upper Jurassic of Africa and the Lower Cretaceous of China. A moderately large pterodactyloid, *Dsungaripterus*, had bony crests along the midline and back of its skull. Its eye sockets were unusually small, its jaws pointed and curved. They may well have been used as forceps to probe for small aquatic organisms, while the blunt, bony projections further back crushed their shells (Fig. 62e). Although considerably smaller, *Phobereptor* had similar bony cranial crests. Its jaws were pointed, but straighter than those of *Dsungaripterus*. Moreover, *Phobereptor* had true teeth (Fig. 62f). It probably fed in a similar manner to that of *Dsungaripterus* (to which it was almost certainly related), but on smaller prey.

6.6.1 Cretaceous Pterosaurs

The Cretaceous period began with the separation of the supercontinents Laurasia and Gondwana (Sects. 2.2, 3.3) and continental drift continued until the end of the Mesozoic Era and then almost stopped. Dinosaurs dominated the land throughout this time but, although birds became increasingly numerous and conquered various terrestrial habitats (Sect. 8.4), pterodactyloids continued their domination of the air – especially over the oceans. They diversified in the Lower Cretaceous, flourished during the middle of the Cretaceous and finally died out in the Upper Cretaceous. In general, they showed an enor-

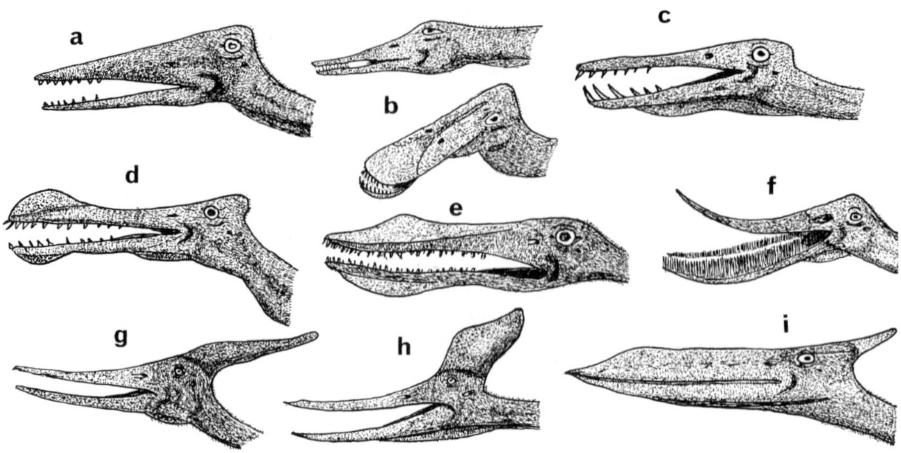

Fig. 63a–i. Heads of Cretaceous pterodactyloid pterosaurs (not to scale). **a** *Ornithocheirus*, **b** *Ornithodesmus*, **c** *Cearadactulus*, **d** *Tropeognathus*, **e** *Anhanguera*, **f** *Pterodaustro* (Lower Cretaceous); **g** *Pterandon ingens*, **h** *P. steinbergi*, **i** *Quetzalcoatlus* (Upper Cretaceous). (Mostly after Wellnhofer 1991)

mous increase in size during this time. Many of the largest forms, such as *Azhdarcho, Titanopteryx, Doratorhynchus, Pteranodon*, and *Quetzalcoatlus* had very long and slender necks with elongated vertebrae.

The skulls of several of these genera have been preserved in the fossil record, and reconstructions of the heads of nine of them are illustrated in Fig. 63. *Ornithocheirus* (Fig. 63a) from the Lower Cretaceous of Europe had long, narrow jaws studded with numerous short, sharp teeth suggesting that it was probably another fish eater. Another rare European Cretaceous pterosaur was *Ornithodesmus* (Fig. 63b). With a wingspan of ca. 5 m, *Ornithodesmus* was twice the size of *Ornithocheirus* (whose wingspan was only ca. 2.5 m). Moreover, it differed from other pterosaurs in that the front ends of its jaws were broad and rounded. This gave them the general appearance of a duck's beak. Unlike that of a duck, however, the front part of the beak of *Ornithodesmus* was armed with short, strong and alternately meshing teeth suggesting that, once again, this pterosaur was piscivorous. So, too, was the South American *Cearadactylus* (Fig. 63c) which, with a wingspan of ca. 5.5 m, was even larger than *Ornithodesmus*. Contemporary with *Ornithodesmus* was *Tropeognathus* (Fig. 63d), a large, toothed and crested pterosaur from Brazil (wingspan ca. 6.2 m), *Santanadactylus, Araripedactylus, Tapejara*, and *Anhanguera* (Fig. 63e). One of the best-known Early Cretaceous pterosaurs, *Anhanguera*, had deep bony crests at the front ends of both upper and lower jaws. These may have served as stabilising devices when the pterosaur was skimming the water for fish as, indeed, may have the jaw crests of *Tropeognathus* (Sect. 6.7.1). The wingspan of *Anhanguera* was ca. 4 m.

Although *Dsungaripterus* had bony knobs in place of true teeth, and some genera were toothless (see below), the majority of pterosaurs possessed fairly simple teeth. An exception was the South American *Pterodaustro* (Fig. 63f). This rare genus had a unique type of dentition. The skull was markedly elongated and the anterior portion curved upwards. The lower jaw was equipped with about 500 bristle-like teeth which could have formed an extremely efficient filtering apparatus. The small organisms filtered from the water were then chopped up by the short, blunt teeth in the upper jaw. The wingspan of *Pterodaustro* (ca. 1.3 m) was only marginally greater than that of *Ctenochasma* (Fig. 62c), but its filtering apparatus was adapted to catch very much smaller animals. It was an ecological equivalent of the flamingos, and probably ate both zooplankton and red algae, as flamingos do. Bakker (1987) even suggested that these pterosaurs may have been pink in colour when alive just like flamingos!

Larger predators tend to capture larger prey unless, like whalebone whales, manta rays (*Manta birostris*), and basking sharks (*Rhincodon typus*), they are so enormous that they have to be filter-feeders (Sect. 5.5). This, of course, is reflected in their dentition: but other factors are involved too – especially in the case of aerial predators. Large teeth are relatively heavy and tend to upset the balance in the air of those predators that possess them. Birds benefit in having horny beaks which are relatively light in weight: the heavy part of their digestive apparatus, the grit in the gizzard, is sited near the centre of lift between the wings. Flying animals with long necks would be especially affected if they bore heavy teeth in their jaws. It is surprising, therefore, that relatively few pterosaurs evolved horny beaks instead of teeth. Those that did so include *Tapejara, Tupuxuara, Nyctosaurus, Pteranodon*, and *Quetzalcoatlus*. The first two were found in the Lower Cretaceous of Brazil, while the remainder date from the Upper Cretaceous. It seems probable, therefore, that toothlessness evolved more than once amongst the pterosaurs, and that the shapes of beaks varied according to their functions, as they do in birds.

Nearly a dozen species of *Pteranodon* (Fig. 53) have so far been discovered. Among the best known are *P. ingens* (Fig. 63g) and *P. sternbergi* (Fig. 63h) from North America. *P. ingens* had a wingspan of ca. 7 m, while that of *P. sternbergi* was up to ca. 9 m. These were all toothless fish-eaters like modern sea birds: fishes and crustaceans have been found in the fossils of their throat sacs (Sect. 6.7.1). Their short, smooth fur was probably kept dry by the use of oil, just as the hair of fruit bats (*Pteropus* spp.) is kept in condition with oil from the neck glands. Moist parts of the body would have been groomed by the legs, which were long and sufficiently flexible to reach almost everywhere, except for the wing claws and the regions adjacent to them, which were within reach of the beak. Their colour was probably dark, as a protection from ultraviolet (UV) light (Bramwell and Whitfield 1974). Finally, the giant *Quetzalcoatlus* (Fig. 63i) probably used its slender, horny beak to probe for molluscs and crustaceans in shallow flood basins (Langston 1981). Fossils of *Quetzalcoatlus* have been found in rock that is almost devoid of fish remains. Burrowing animals were, however, present and the pterodactyl probably fed on these. Fossilised logs in the same strata suggest that periodic flooding took place, possibly resulting from a monsoon type of climate.

Wellnhofer (1991) has described many other genera of pterosaurs, often based on quite fragmentary fossil evidence, but superbly illustrated by John Sibbick. Clearly, the pterodactyloids enjoyed worldwide distribution from the beginning of their evolution.

6.7 Life Styles

The pterosaurs were the first vertebrates fully to conquer the air. Although many fossilised examples have been collected, much of what is known about their modes of life is based on inference – as we have seen in the case of their locomotion on the ground (Sect. 6.4).

6.7.1 Feeding

Apart from a few small insectivorous forms such as *Peteinosaurus*, *Batrachognathus* and *Anurognathus*, most of the pterosaurs so far discovered have been specialised fish-eaters. Their long jaws and forwardly directed front teeth would have enabled them rapidly to seize slippery fish and to prevent them from escaping. In several cases, the remains of partly digested fishes have been observed in the fossils of pterosaurs. As already mentioned, many pterosaurs probably fished on the wing with lower jaws hanging in the water, as do skimmers (*Rhynchops* spp.) today. The tips of their jaws had horny beaks which have not been preserved in their fossils. Had they landed on the water surface whilst fishing, they could easily have taken off again by facing the wind, spreading their wings (Sect. 6.3) and paddling their feet, as do albatrosses (*Diomedea* spp.). The jaw articulation of the pterosaurs shows that they could have opened their mouths very widely. When they did this, their jaws would have splayed out, thereby increasing the opening to the throat pouch in which food was temporarily stored – as it is by pelicans (*Pelicanus* spp.). Perhaps food was pre-digested in the pouch before being swallowed or brought back to feed the nestling young (Sect. 6.7.2; Wellnhofer 1991).

To conclude, the smaller pterosaurs were mainly insectivorous while the larger species were largely piscivorous. Among the latter, larger forms usually fed upon larger fishes, but there were many genera of roughly the same size in which the teeth, paradoxically, were markedly different. However, there are more ways than one of solving a problem, and one type of piscivorous dentition may be quite as efficient as another. When two genera of similarly sized pterosaurs but with different dentition existed contemporaneously it could well reflect different prey or different ancestry or both. And, if the two had evolved at different times from different ancestors, the apparent paradox would be simply explained.

In the African savannah, most of the terrestrial predators are also scavengers. Nevertheless, they do not eat anything like the amount of flesh that is consumed by vultures. Soaring at great heights, vultures have a much better chance of spotting the corpses of animals that have died from wounds or disease than terrestrial scavengers do. When one vulture drops to the ground it is seen by others from far away. These follow it to its destination attracting, in turn, other vultures even more distant. The same could have applied in the case of pterosaurs, but there is no evidence that scavenging provided much of their diet – even in the case of *Quetzalcoatlus* which, it has been suggested, might possibly have fed on the carcasses of dinosaurs, using its slender beak to probe into their bodies. Nor are there any records of fruit-eating pterodactyls although, when the first higher flowering plants evolved, some species would almost certainly have been responsible for distributing their seeds. Birds were relatively scarce during the Late Cretaceous period (Sect. 6.5) but frugivorous pterosaurs probably lived inland in higher regions subject to weathering and erosion. Here, the chances of fossilisation would have been small (Fleming and Lips 1991; Wellnhofer 1991). Fleming and Lips (1991) based their argument on three premises: (1) pterosaurs had a metabolism similar to that of birds; (2) their adaptive radiation was more extensive than is indicated by the fossil record; and (3) birds and bats evolved a frugivorous diet, and pterosaurs probably did so too (see, however, Sect. 6.8).

6.7.2 Reproduction

It is not yet known for certain whether the pterosaurs gave birth to living young or laid eggs, but it seems more probable that they were oviparous. Fossils of eggs, which could easily have been those of pterosaurs, have been discovered on several occasions. In the case of *Quetzalcoatlus*, fossils of eggs have been found at the same site in Texas as have fossils of the adults. As yet, however, there is no positive proof that the two were associated. So far, no fossils of adult pterosaurs have been found with either eggs or embryos in their bodies. Indeed, there is no indication of the gender of most individuals although in the case of certain species of *Rhamphorhynchus* there are sufficient fossils for statistical confirmation of the fact that two forms are present. One, probably representing the males, has larger skulls and wings than the other that may consist of females (Wellnhofer 1991). *Quetzalcoatlus northropi* (wingspan ca. 12 m) is known only from wing fragments, but bones of another form with a wingspan ca. 7 m have also been found. These bones could represent another species, the opposite sex, or may perhaps have belonged to immature individuals (Langston 1981). Even larger than *Q. northropi*, *Hatzogopteryx* had a wingspan in excess of 12 m and a skull ca. 1.5 m in length.

Among birds, the males are usually larger than the females. There are exceptions, however, as in the case of birds of prey (Falconiformes). In these, the females are markedly larger than the males, and the two sexes may even hunt different quarry. Not only size differences, but also sexual dimorphism occurs in birds. For instance, the bill and casque of the male black-casqued hornbill (*Caratogymna atrata*) are very much larger than those of the female. Differences in size and dimorphism may well have been present among pterosaurs. Indeed, both have been assumed in the case of *Pteranodon* in some species of which it is possible to distinguish a larger form with a big cranial crest from a smaller form. This had a smaller cranial crest and a wider pelvic canal so, presumably, it is the female (Bennett 1987).

The pelvis of pterosaurs was quite different from that of birds, so the eggs must have been rather small and, if the young were born alive, they would have been very immature at birth. Either way, parental care must almost certainly have taken place. With all the emphasis on the reduction of weight in pterosaurs, it would scarcely have been possible for the females to have carried more than one egg or fully formed young at a time. Nests must therefore have been built and the offspring reared in them. Masses of fossilised conifer needles associated with the bones of *Quetzalcoatlus* may be the remains of such nests.

Since they were endothermal and hairy, it is quite possible that the pterosaurs incubated their eggs as birds do. Like birds, they would have paired for at least a season, if not for life, and the males might have taken their turn at incubating or have fed the females from their pelican-like pouches. Alternatively, pterosaurs might have placed their nests in secure places where they were

Fig. 64. *Ornithocheirus* (Pterodactyloidea; Lower Cretaceous; wingspan ca. 2.5 m) feeding its young. (After Wellnhofer 1991)

warmed by the sun and have taken no part in incubating the eggs. The hatchlings would, however, have been extremely small and the parents undoubtedly had to feed them. Figure 64 shows what *Ornithocheirus* might have looked like whilst feeding a fairly large youngster.

Pteranodon, like many other pterosaurs, probably lived in colonies where suitable roosting places were available. Each individual would have needed sufficient space to spread its wings in order to be able to take off into the air. Mating took place in the hanging position, as it does in bats, and the young would have been much safer in a colonial nesting site than they would have been were the animals solitary in habit (Bramwell and Whitfield 1974). Similar considerations may well have applied to many other pterosaurs, but little evidence is available on the subject.

Wellnhofer (1970, in Wellnhofer 1991) was able to reconstruct the growth pattern of *Pterodactylus* on the basis of statistical investigations. In the case of *P. kochi*, the smallest individual had a wingspan of only 19 cm, while the largest had one of 67.5 cm. A continuous growth series was assembled between these two extremes and it was shown that the skull grew more quickly in young than in older individuals. The eye socket was relatively larger in younger individuals, as it is in crocodiles and birds. Many other parts of the skeleton also showed differential growth rates.

6.7.3 Enemies

Pterodactyls were skinny creatures and would not have been very attractive to predators. Their enemies would doubtless have included carnivorous dinosaurs on land, and marine reptiles especially mosasaurs in the sea. On land they would have launched themselves from their roosting places to escape. On the water, if they had time, they would also have taken flight. If flight was impossible, they might have used their relatively large wings and jaws in deimatic threat. (Intimidating or startling behaviour is called 'deimatic' after the Greek word meaning 'to frighten'.) Nevertheless, like birds and bats, pterosaurs would have had few defences whilst on the ground, and their natural response to enemies must inevitably have been to take to flight.

6.8 Extinction

Extinction, resulting from competition and environmental change, is the inevitable fate of every species and taxon of living organisms. Only when mass extinctions occurred (Table 3) is there any call for explanation or comment. When many groups of organisms disappeared simultaneously within a relatively short space of time, the change recorded in the fossil records appears to have taken place so rapidly that some people are tempted to assume that a catastrophic event must have been responsible. The Rhamphorhynchoidea reached their peak of diversification shortly before their extinction at the end of the Jurassic period, while a few of the Pterodactyloidea persisted almost until the end of the Cretaceous. Toothless pterosaurs which dominated during the Upper Cretaceous appear in the fossil record only as isolated individuals – the toothed forms had already disappeared before this. Pterosaur history therefore included two periods during which they flourished – the Late Jurassic and the middle of the Cretaceous and three extinction events. These were at the end of the Jurassic, in the Early Cretaceous, and towards the end of the Mesozoic Era. There were, however, no mass extinctions at the times when most groups died out.

Bramwell and Whitfield (1974) speculated that the extinction of *Pteranodon* could have been caused by climate change, and particularly in average wind speed towards the end of the Cretaceous. This would also have applied to *Quetzalcoatlus, Titanopteryx*, and *Azhdarcho.* An increase of only 5 m s^{-1} would have been enough to make conditions impossible for the giant pterosaurs. Such a change could have been caused by global cooling, accompanied by the development of clear temperature differences between the equator and the poles. As Wellnhofer (1991, incorporated in Norman and Wellnhofer 2000) argued, longer periods of the year with higher wind speeds, during which the large pterosaurs were unable to fly, would have reduced their numbers to such an extent that they sank below the critical level at which survival was possible.

As the landmasses of Eurasia drifted apart, the marine currents which influence world climatic conditions would have become more like those which obtain today. Consequently, the adaptive process which led to the evolution of the great pterosaurs would have led to their demise. In order to survive, they would have had to become smaller, develop heavier bones and acquire flapping flight once again. According to Unwin (1988), pterosaurs and birds showed differential survival patterns of relative diversity during the Mesozoic. This conforms to a double-wedge pattern typical of the differential survival model of faunal replacements. Birds which began to compete with the pterosaurs during the Lower Cretaceous (Sect. 6.7.1) were better adapted to the changing conditions, and during the Upper Cretaceous would have competed more strongly with them – even over the oceans that were their last stronghold. However, all this is very speculative. Nothing is known of weather patterns in those days and there is no real evidence either for, or against, competition between pterosaurs and birds. Certainly, there is little evidence to suggest that the pterosaurs radiated as birds and bats have. In fact, they appear to be the analogues of the sea birds today (Naish and Martill 2003).

6.9 Summary

Palaeontologists are generally in agreement that the pterosaurs would have flown almost, if not quite, as well as most modern birds. On the other hand, their locomotion on the ground, about which there is still some controversy, was almost certainly awkward and ungainly. Their flight may indeed have been like that of birds, but their locomotion and roosting behaviour were probably more like those of bats than of birds. Most of the smaller species were insectivorous; larger forms were piscivorous. Their dentition was adapted to their prey. They nested colonially and fed their nestling young until these were able to care for themselves.

7 Terrestrial Mesozoic Reptiles: Size, Locomotion and Thermal Physiology

7.1 Introduction

The Mesozoic Era may indeed have been the 'age of dinosaurs' but, as we have seen (Chap. 2), there were many other dramatic reptiles inhabiting the land in those days – not to mention the amphibious (Chap. 4), marine (Chap. 5), and aerial (Chap. 6) forms we have been discussing. Terrestrial Mesozoic reptiles included tortoises and turtles (Anapsida: Testudines), the mammal-like reptiles (Synapsida: Therapsida) and various Diapsida. Among the latter were the sphenodontians, the squamata – lizards and snakes – and the archosaurs. These will be discussed in the following chapters. The classification of Mesozoic reptiles (Table 2, Fig. 1) has been clearly elucidated and explained among others by Benton (1990a, 2004) and Benton and Harper (1997). Although the archosaurs were the main archosauromorph diapsids to emerge in the Triassic, there were other land-living groups too. The dinosaurs first appeared in the Upper Triassic (230–225 mya) and, perhaps surprisingly, were at first bipedal (Sect. 7.3).

7.2 Quadrupedal Locomotion

When reptiles first evolved from amphibians (Sect. 1.3), they inherited the short, stumpy legs of their ancestors. Modern newts and salamanders (Urodela) as well as modern reptiles such as lizards with elongated bodies and small laterally projecting legs show the most primitive form of tetrapod locomotion. If they are frightened and need to move exceptionally fast, urodeles and lizards usually wriggle along with the belly resting on the substrate, as though they were swimming, with their legs scarcely touching the ground. This eel-like movement is similar to the serpentine locomotion of a snake (Sect. 4.2, Fig. 16). When moving deliberately, however, both amphibians and quadrupedal reptiles carry their weight clear of the ground. The movement of one forelimb is followed by that of the hindlimb on the opposite side of the body. The forelimb of that side is then moved forward, followed by the rear leg on the opposite side. When these movements are accelerated, however, and the animals begin to trot, the limbs opposite each other tend to be moved simultaneously and, during most of each strike, only two feet are on the ground at the same time (Gray 1968, summarised in Cloudsley-Thompson 1999).

Correlated with the basic similarity of movement between small, elongated reptiles and urodeles, there is a general similarity in the plans of their skeletons. Nevertheless, certain advanced features are characteristic of all reptiles. The head is usually carried above the level of the ground, and the neck is well developed. The first two cervical vertebrae are modified to form the atlas and axis bones, and all the vertebrae articulate with one another by a system of interlocking processes much more elaborate than anything found in the Amphibia.

The vertebral column provides a stiff but flexible support for the body, and the legs are connected to it by the pectoral and pelvic limb girdles. In the case of amphibians and most reptiles, excluding dinosaurs (Sect. 7.3), the limbs are held out from the sides of the body and the animal adopts a sprawling posture. To maintain this position involves the expenditure of a considerable amount of energy, especially in the case of the heavier reptiles such as large turtles and crocodilians – which therefore spend most of the time with their bodies resting on the ground. Legs do more than merely support the body, however: they also provide locomotion. These two functions do not operate synergistically. Weight bearing is provided for most efficiently by adopting an erect posture with the legs articulated below the limb girdles. Speedy movement, on the other hand, is better achieved by having the limbs inclined outwards, as in lizards and other small reptiles. In practice, a compromise between these two objectives is reached.

Locomotor performance in an erect posture is improved when the leg segments are inclined towards one another. Thus, the humerus slopes forwards, the radius and ulna backwards; while the femur slopes forwards and the tibia and fibula slope backwards. The inclination of the limb segments is maintained by tension in the muscles and ligaments. This provides for a natural springing action that allows for the storage and release of strain energy. Speed is increased by having the limb segments move relatively to each other, because the speed of the entire leg represents the sum of speeds of its component parts (McGowan 1991). Quadrupedality has been independently derived from bipedal ancestors in the sauropods (Saurischia) and ankylosaurs, stegosaurs, ceratopsians, hadrosaurs, and iguanodonts (Ornithischia).

7.3 Bipedal Locomotion

The earliest dinosaurs did not exceed 2 m in length and were lightly built. They were bipedal, as already stated, and their hind legs were considerably longer than their forelimbs – as were those of ancestral crocodilians (Sect. 4.6.3). In dinosaurs, a bipedal stance and parasagittal gait were both primitive and obligatory. (Parasagittal means parallel to an animal's midline.) Dinosaurs would certainly not have been able to sprawl (Padian 1997; Parrish 1997). Although the earlier dinosaurs (Fig. 65) – like most of their archosaurian predecessors and contemporaries – were bipedal and had erect stances, the tibia and

Fig. 65. *Plateosaurus* (Sauropodomorpha; Upper Triassic; length ca. 7 m)

fibula were usually longer in dinosaurs than was the femur. In contrast, the femur was longer than the tibia and fibula in most of the non-dinosaurian archosaurs. Dinosaurs also held the metatarsus off the ground in a digitigrade stance, whereas the other archosaurs retained the primitive plantigrade condition in which the entire foot was placed on the ground during locomotion (Parrish 1997).

Plateosaurus, the genus illustrated in Fig. 65, is the best known of the Sauropodomorpha or prosauropods (Sect. 10.3) because numerous well-preserved fossils have been unearthed in Triassic sandstones throughout Europe. In some places, groups of complete individuals have been found together, suggesting that the animals may well have been gregarious and travelled together in herds throughout the desert landscape, searching for new feeding grounds. Alternatively, of course, the bodies of animals that had died previously might have been washed together in periodic flash floods or have been piled up at the ends of wadis. *Plateosaurus* was a large reptile. Its tail, towards which the intestines had extended (see below) was nearly as long as its body.

The tail would have had a major effect on locomotion in bipedal dinosaurs, especially when they were running. The trunk was erect and, when one foot was set down in front of the centre of gravity, it would not have been lifted until it was behind the centre of gravity, as in the case of human beings. A stiff tail, useful also as a weapon (Sect. 9.2.2) would have been better for running than would a springy tail. Had bipedal dinosaurs jumped, as kangaroos do, springy tails would doubtless have evolved, as they have in kangaroos, to reduce the rocking movements of the trunk (Alexander 1989).

A bipedal stance is more favourable for carnivores than for herbivorous animals. This is due to the fact that the digestion of vegetable matter necessitates relatively larger intestines which have to be accommodated in larger, heavier bodies in front of the hips. These, in turn, cause problems of balance. Selection was consequently pulling in two opposite directions. Shorter torsos provided better balance for running, while longer bodies were needed for digesting leaves (Bakker 1987). The way in which the dilemma was solved evolutionarily was by bending the pubic bones backward from their attachment to the other hip hones so that the intestines could lengthen, while the hip joint remained as it had been. The pubic bones now slanted downward and backward instead of directly down, and the intestinal canal could be prolonged from the belly to the base of the tail (Galton 1970a). Bipedal balance was thereby actually improved by the increased weight of the body.

7.4 Weight and Size Limits

The body proportions of animals are affected by a number of considerations, probably the most important of which are the ratios of size to mass. If two individuals were to have exactly the same shape, but one was twice as long as the other, all other linear dimensions would be double those of the smaller individual. The surface area would, however, be the square of the surface area of the smaller individual, while its mass would be cubed. Thus, if a reptile 3 m in length with a surface area of 4 m^2 weighed 150 kg, a similarly shaped individual 6 m long would have a surface area of 16 m^2 and would weigh 3.375 tonnes. This huge increase in mass relative to linear dimensions invokes one of the many problems involved in scaling (Schmidt-Nielsen 1984; Alexander 1997b). Some consequences of size are not closely related to the laws of physics. For instance, the effectiveness of an eye depends upon is absolute size, not on the size of its possessor. Consequently, larger animals tend to have smaller eyes relatively to their body mass. The same applies to brains. There is indeed a rough correlation between the size of its brain and the intelligence of an animal. Intelligence, however, is proportional more to the surface area of the brain than to its weight, so that mere percentage brain-weight is no criterion of intelligence. The dinosaurs may not have been quite so slow, unresponsive and silly as is sometimes thought! (see Buchholtz 1997).

Greater weight involves considerably stronger bones. When an animal stands on land, stresses are set up in the bones and muscles that support its weight. Whereas the cross-sectional areas of bones and muscles in isometric animals are proportional to (body weight)$^{0.67}$, the stresses due to weight are proportional to (body weight)$^{0.33}$. Consequently, since large animals are constructed of the same materials as their smaller relatives, their bones and muscles must be disproportionately thick in order to withstand the same stresses,

Fig. 66. *Diplodocus* (Sauropoda; Upper Jurassic; length ca. 25 m). (Cloudsley-Thompson 1994)

or they must stand and walk in such a way as to minimise stresses, or both. This problem does not arise, of course, in the case of aquatic animals whose weight is supported by buoyancy (Alexander 1971). The ability of bone to resist compression, the relationship between mass and surface area, as well as various physiological problems, including thermal ones (Sect. 7.5), also affect larger animals (Cloudsley-Thompson 1977).

Many Mesozoic reptiles were extremely large and heavy. The Diplodocidae ('double beams'), for example, were enormous plant eaters and included some of the longest of all dinosaurs. Among the best-known genera are *Apatosaurus* (= *Brontosaurus*), *Atlantosaurus, Brachiosaurus, Diplodocus* (Fig. 66), and *Seismosaurus* the 'earth shaker' (Gillette 1994). Despite the fact that they were relatively lightly built – their backbones contained deep weight-reducing hollows – they were, nevertheless, extremely heavy. They maintained an erect posture and had weight-bearing or graviportal limbs, as do elephants and rhinoceroses today.

The limb bones of *Apatosaurus* were massive. Most of the weight would have fallen on the hindlimbs so, for an individual weighing about 30 tonnes, each time one rear leg moved forward, the other must have carried about 20 tonnes (Fig. 67). From this shape, according to McNeill Alexander (1971), they would have been most likely to break at *XX* and *XY* in the figure. Alexander calculated

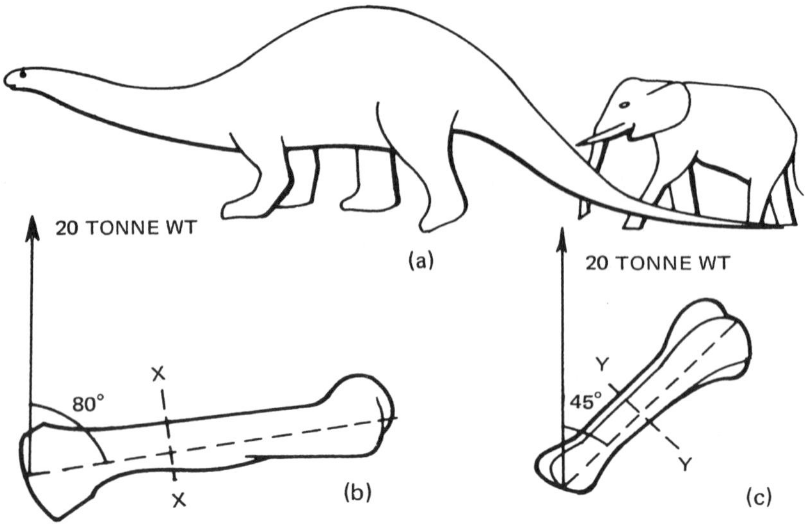

Fig. 67. a An *Apatosaurus* as it probably appeared in life, with a large African elephant drawn to the same scale; **b** the femur; **c** the tibia and fibula of *Apatosaurus* showing forces that would just break them. (Cloudsley-Thompson 1978 after Alexander 1971)

from the tensile strength of bone that they would just be broken at these places by forces of 20 tonnes. If the muscles were strong enough, the animal could have taken steps at up to about 45° to the vertical with the femur almost horizontal. So the bones were quite strong enough, provided that excessively long steps were not taken and the dinosaur did not stumble. Fossil footprints show that the steps taken were 2.4 m in length, which is quite modest for an animal with hind legs more than 3 m long. No doubt larger genera would have taken relatively smaller steps. The larger they were, the less agile they must have been, just as elephants are less agile than smaller mammals (Alexander 1971).

Thulborn (1990) cited the following calculated speeds: *Brachiosaurus* (78,258 kg) 18 km/h; *Apatosaurus* (32,418 kg) 24 km/h; *Diplodocus* (10,562 kg) 12 km/h; *Tyrannosaurus* (6,895 kg) 23 km/h; and *Triceratops* (847 kg) 26 km/h. His figures and those based on the work of T. Garland were comparable, but those of R.T. Bakker were more than twice as high. A more recent analysis by Hutchinson and Garcia (2002), based on estimates of the extensor muscle mass needed to support its body weight compared with what modern birds and cursorial mammals possess, suggests that *Tyrannosaurus* (6,000 kg) could not have exceeded speeds of 29–36 km/h (8–10 m/s). (Earlier methods for estimating the speeds of dinosaurs are given below.)

At one time it was thought that the large sauropod dinosaurs had exceeded the weight limit, and must have spent all their lives with their weights supported by water. The discovery of the footprints of adult and baby dinosaurs beside one another along the Paluxy River in Texas has proved that *Apato-*

Fig. 68. *Brachiosaurus* (Sauropoda; Upper Jurassic; length ca. 25 m)

saurus excelsus (ca. 35 tonnes) must have been terrestrial (Bird 1944). Moreover, to have used their long necks as snorkels so that they could breathe while standing in deep water would have been a difficult process. In the case of *Diplodocus*, the lungs might have been 6 m below the surface. To have expanded its lungs against the pressure of water at this depth would have needed enormous chest muscles. It seems probable, therefore, that the largest of the sauropods would have kept clear of marshy places where they might have got bogged down (Alexander 1985, 1989). Another of the reasons why the larger dinosaurs – such as *Brachiosaurus brancai* (Fig. 68), which may have weighed about 80 tonnes – were thought to have been aquatic is that their external nares were situated high on the skull, extending posteriorly to just in front of the orbit. This would have enabled their possessors to breathe with the body entirely submerged, as crocodiles do. However, the living reptiles with a narial structure most closely resembling that of the apatosaurs are terrestrial – ground iguanas (*Cyclura* spp.) and desert monitors (*Varanus griseus*). Again, in most truly aquatic reptiles – crocodilians, phytosaurs (Fig. 29), *Champsosaurus* (Fig. 30), and so on – the tail is a powerful propulsive organ. In contrast, the tail of the apatosaurs was slender, ending in a whiplash like that of many terrestrial lizards, past and present.

The vertebral column of the apatosaurs had several features associated with the support of the body on dry ground, and its rib cage was very deep, resembling that of elephants rather than of hippos. The limbs, too, were elephant-like in their proportions as we have seen, while the diversity of the sauropods gen-

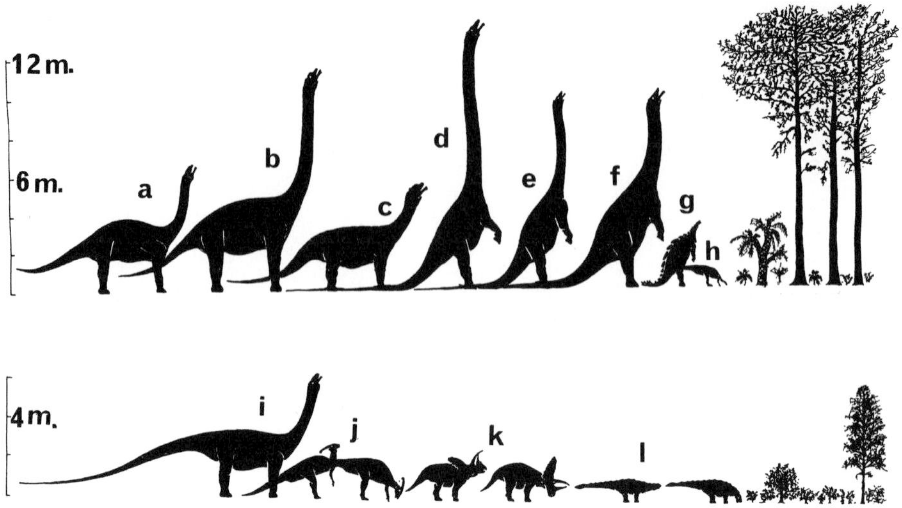

Fig. 69a–l. Possible postures of some dinosaurs when browsing on foliage. *Above* Upper Jurassic. **a** *Haplocanthosaurus*, **b** *Brachiosaurus* (not a large one), **c** *Camarasaurus*, **d** *Barosaurus*, **e** *Diplodocus*, **f** *Apatosaurus*, **g** *Stegosaurus*, **h** *Camptosaurus*. *Below* Upper Cretaceous. **i** *Alamosaurus*, **j** *Parasaurolophus*, **k** *Triceratops*, **l** *Ankylosaurus*. The Jurassic apatosaur and stegosaur are feeding on cycads and conifers; the Cretaceous apatosaur and low-feeding beaked dinosaurs are browsing on angiosperms. (Adapted from Bakker 1987)

erally also argues for terrestrial habits (Fig. 69). For these and other reasons, it is now accepted that the deep chest, long limbs, and long necks of all the sauropod dinosaurs were adaptations for browsing on land, and were not related to an amphibious or aquatic existence (Fig. 70; Bakker 1971b).

For estimating sizes and size limits, the weights of dinosaurs are more important than are their heights or lengths. Unfortunately, there is considerable uncertainty about calculating the masses of dinosaurs. For example, E.H. Colbert (1962a) estimated the weight of *B. brancai* as 78 tonnes – using the model method. In 1989, Alexander's estimate was 47 tonnes, but the previous year G.S. Paul had calculated that it was only 32 tonnes. The reason for these differences is that the various authors based their calculations on different fossil individuals. Unmounted limb bones of an *Apatosaurus* from Tanzania are about 30% larger than those on which Colbert based his calculations! Furthermore, some authors used models in which the dinosaur represented was a bulky animal; others used skinny models. To add to the uncertainty, the densities of dinosaurs are not known in the first place. It also has to be assumed that dinosaur bone was about as strong as the bones of modern birds and mammals. This is probably the case, but we cannot know for certain (Alexander 1989, 1997a).

Bakker (1978, 1987) claimed that not only did the sauropod dinosaurs use their long necks to feed from the tops of trees as giraffes do, but that some of them also reared up on their hind legs propped by their tails (Fig. 68). As

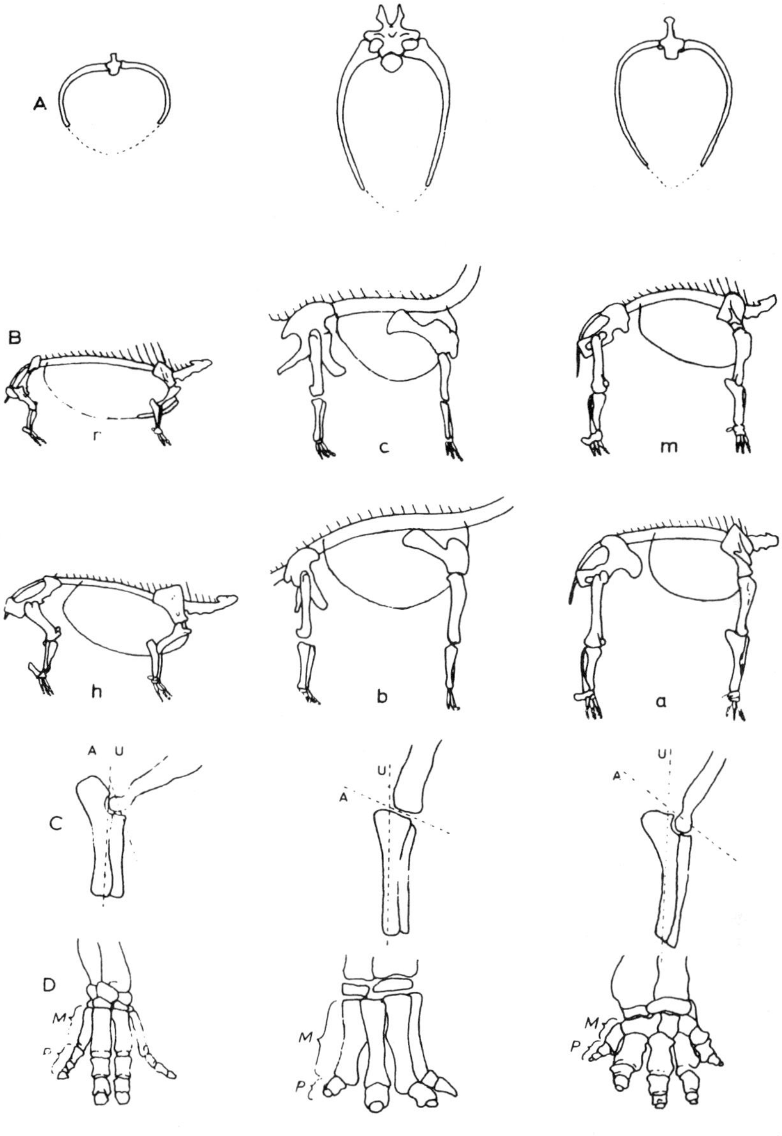

Fig. 70A–D. Comparison of hippo (*left*) apatosaur (*centre*) and elephant (*right*). **A** Cross section of thorax of hippo, apatosaur (*Diplodocus*) and elephant (*Loxodonta).* **B** Outlines of right side of skeletons drawn to same acetabulum-to-shoulder length; *r* hippo-like rhino (*Teleoceras*); *h* hippo (*Hippopotamus*); *c* apatosaur (*Camarasaurus*); *b* apatosaur; *m* elephant *(Mastodon); a* elephant (*Archidiskodon*). Full length of apatosaur (*Brachiosaurus*) necks not shown. **C** Section through elbow to show orientation of articular surfaces. *Left* Hippo (*Hippopotamus*); *centre* apatosaur (*Camarasaurus*); *right* elephant (*Elephas*). *A* Axis of elbow facets on humerus. *U* axis of facets on the radium and ulna. **D** right forefeet of hippo (*Hippopotamus*), apatosaur (*Diplodocus*), elephant (*Mastodon);* *M* metacarpals; *P* phalanges. (Cloudsley-Thompson 1978 after Bakker 1971b)

Fig. 71. *Stegosaurus* (Ornithischia; Upper Jurassic; length ca. 6 m). (Cloudsley-Thompson 1994)

Beverly Halstead pointed out in a review of Bakker's, *The Dinosaur Heresies* (1987), the evolution of dinosaur stance and gait was worked out by Alan Charig as long ago as 1965, but is presented without acknowledgement. In order to test the validity of Bakker's claims, Alexander (1985) tried to estimate the positions of the centres of gravity of sauropods and some other dinosaurs to see if it would have been possible for them to have reared up on their hind legs. He argued that the densities of dinosaurs were probably nearer to 1,000 kg m^{-3} than to 900 kg m^{-3}, as Colbert (1962a) had assumed, and he concluded that both *Diplodocus* (Fig. 66) and *Stegosaurus* (Fig. 71) supported about 80% of their weight on their hind feet, and only about 20% on their forefeet. They would therefore both have been capable of rearing up on their hind legs. The trunk of *Triceratops* (Fig. 72) probably tended to sag and would have required tension in the hypaxial muscles to prevent this. *Diplodocus*, on the other hand, had no tendency to sag because of the counterbalancing weight of the neck and tail. The neck was presumably supported by a strong nuchal ligament running through the forked neural spines of the cervical and dorsal vertebrae. The leg bones of *Apatosaurus* were probably strong enough to have permitted the degree of agility shown by modern elephants. Those of *Diplodocus* were too weak for running while those of *Triceratops* were strong enough for even greater agility than that of elephants. The pressures on the soles of the feet of large biped dinosaurs were about the same as those of cattle, but higher pressures acted on the feet of sauropods.

Dodson (1990) agreed that sauropods may have reared up on their hind legs on special occasions such as when fighting or reaching for the highest leaves, but he thought it unlikely that they would have reared up very often because, if the head was held high above the heart, an extremely high blood pressure would

Fig. 72. *Triceratops* (Ornithischia; Upper Cretaceous; length ca. 9 m). (Drawing by Anne Cloudsley)

have been necessary to maintain an adequate supply of blood to the brain. Clearly, the cardiovascular pressures in sauropods would have been enormous. Hohnke (1973) pointed out that a specimen of *Brachiosaurus* (Fig. 68) in the Berlin Museum showed a vertical blood column of about 6.5 m above the heart. This is about twice that of a giraffe, and would have required a ventricular pressure of over 500 mm of mercury (Hg) just to support the column! (Alexander 1997a). Alternatively, the apatosaurs might not have raised their heads high or, if they did, they might have endured a temporary cessation of blood flow to the brain (Seymour 1976).

7.5 Thermal Physiology

The thermal physiology of the dinosaurs and of other large reptiles has been the subject of animated discussion, not least because it is relevant to their extinction at the end of the Cretaceous period (Chap. 12). Temperature regulation in animals is either behavioural, autonomic (self-governing) or, more usually, a combination of the two. Like most day-active animals, Mesozoic reptiles would have basked in the sun when they were cold and sought the shade if they were too hot. Nocturnal species, on the other hand, would have taken shelter during the daylight hours and emerged at dusk when the air was cooler. The Nile crocodile (*Crocodylus niloticus)*, for example, maintains a relatively constant body temperature by spending the night in water, sun basking in the morning and evening, and retreating into the water or shade during the heat of the day (Cott 1961).

The rate of metabolism and, therefore, of heat production increases with body size in most of the larger homeothermic animals, but not in proportion to it. When the logarithm of metabolic rate (MR) is plotted against the logarithm of body weight (BW), however, a straight line is obtained. The slope of this line indicates that MR is a function of $BW^{2/3}$. Since the surface area of a sphere varies with $mass^{2/3}$, as explained above (Sect. 7.4), this relation is an indication of the influence of surface area on MR. The maintenance of a body temperature above that of the environment and at more or less the same level irrespective of size requires that the rate of heat production per unit mass of tissue must be proportional to the rate of heat loss. There is no way of estimating the metabolic rates of Mesozoic reptiles but, if the larger species had metabolic rates as high as those of extant birds and mammals, their surface areas might well have been insufficient to allow for adequate cooling of their bodies. Using tortoises (*Testudo graeca*) as model dinosaurs, David Butt and the present author made simultaneous measurements of metabolism and transpiration at different ambient temperatures. Up to about 24 °C the metabolic heat gained and heat lost by evaporation kept approximately in pace with one another. At temperatures above this, however, there was an increasing excess of heat production over heat loss (Cloudsley-Thompson 1978, 2001). It seems inconceivable that the metabolic rates of dinosaurs should not have been at least as high as those of a slow, solid tortoise!

The main points at issue are whether the large Mesozoic reptiles, in particular the dinosaurs, maintained high and more or less constant body temperatures endothermically, as modern mammals and birds do, or whether they were heterothermal – with variable body temperatures – and had lower metabolic rates like those of extant reptiles.

7.5.1 Behavioural Thermoregulation

Small extant reptiles lack effective insulation, but exercise control over their body temperatures through a combination of behavioural and physiological body mechanisms. They are ectothermal – most of their body heat is obtained from environmental sources, and comparatively little is generated by metabolism. They are also poikilothermic and their body temperatures are not controlled precisely like those of homeothermic endotherms, most of whose body heat is generated metabolically. Thermoregulation by behavioural means is an excellent strategy for small animals that have a large surface-to-volume ratio, but it is a mechanism far less readily applicable to large animals, as we shall see.

The most important source of environmental heat is solar radiation, and most day-active reptiles today divide their time between basking in the sun and cooling themselves in the shade. Such behaviour would be far less practicable for a large animal such as a dinosaur, and there are additional complications as well. Colbert et al. (1946) found experimentally that, under similar conditions, whereas a very small alligator (weighing about 50 g) needed 7.5 min to warm from 27 to 32 °C, a large alligator (weighing about 13,000 g) required 30 min to heat up from 28 to 32 °C. In other words, the small alligator heated at a rate of 1 °C per 1.5 min, the larger at a rate of 1 °C per 7.5 min. When these results were extrapolated, it was found that a reptile the size of a medium-sized dinosaur weighing about 9,000,000 g (9 tonnes) could be expected to warm up at a rate of only1 °C per 9.6 h (5,760 min)!

Of course, this medium-sized dinosaur, having a small surface-to-volume ratio, would retain its heat overnight. Seasonal changes would, however, present a problem, and it might experience considerable difficulty in finding a cool retreat if it got too hot. Indeed, it would be dangerous for its body temperature to approach the critical maximum because of the amount of time needed for it to cool down. Again, the juvenile dinosaurs might also have been susceptible to unusually high temperature over short periods which would not have affected the adults adversely. Colbert et al. (1946) argued that upland dinosaurs might have reacted differently from swamp-living forms, from which they differed in size and shape. They would have spent much of the time in shade, whereas amphibious forms would have been able to seek relief from the sun's heat, as crocodiles do. Bipedal forms may have exercised some thermal control by facing the sun except during the early hours of the morning and in late afternoon. Alligators in a bipedal posture were found to show the slowest rate of rise in temperature when exposed to the sun: alligators flat on the ground showed the highest rate of rise, while animals held in a quadrupedal pose heated up at a rate intermediate between that found in the bipedal posture and that experienced when flat on the ground.

It is conceivable that the tremendous size of many of the archosaurs, as well as the prevailing bipedal posture, were the direct results of thermal limitation. The erect posture presents a smaller surface to insolation and removes the body from contact with the ground, while the great volume requires a longer period to heat as well as a longer period to cool off during the night (Cowles 1940). In fact, as shown by the calculations of Spotila et al. (1973), changes in environmental temperature were probably damped out over periods of several days. Axelrod and Bailey (1968) believed that the mean annual temperature in equatorial regions at the end of the Cretaceous period was probably not much above 27 ° C (80 °F). Moreover, they pointed out that most large reptiles today escape excessive heat by submergence. It is also believed that many of the larger dinosaurs were to some extent amphibious, and that others, such as the herbivorous trachodonts, were at least partially so, to judge from their webbed feet (see Cloudsley-Thompson 1978).

7.5.2
Thermoregulatory Structures

The 'sail' of the extinct Pelycosauria may have been a device for absorbing solar heat in the Permian mornings and for radiating it again when the animals became overheated (Sect. 2.5); the numerous grooves in the spines suggest that the skin covering it may have been well supplied with blood vessels (Romer 1948; Rodbard 1949; Bellairs and Attridge 1975). This hypothesis has recently been supported by the calculations of Bramwell and Fellgett (1973), which showed not only that the sail of *Dimetrodon grandis* could have enhanced heat gain, but it would also have been capable of radiating heat. They stated: "Faster attainment of the activity temperature in the morning would have been an obvious advantage to a carnivorous reptile feeding on other large poikilothermic animals. *Dimetrodon grandis* would have been able to reach an active state and attack prey while they were still torpid or sluggish. During the hot part of the day the sail would have acted to radiate away excess heat; this effect could be an important adaptation in a reptile which was too large to seek shade behind small stones or in rock crevices in the manner of small living reptiles. In the evening *Dimetrodon* could have gained extra activity time, in comparison with a small sail-less pelycosaur of the same A/W (area to weight) ratio, by restricting blood flow to the sail. The sail could therefore prolong the total time in which *Dimetrodon* could be active in any 24 hr."

"Our conclusions are reinforced if, as seems probable, the controlling adaptations had become highly refined. In addition to adjustments of blood supply, the degree of blackening may have been under nervous or hormonal control. Some lizards which are black when basking to catch the maximum radiant energy can change to white when oriented head-on to the sun and emitting heat. It is possible that *Dimetrodon* (or indeed living reptiles) may be black in the infra-red to radiate heat more effectively, while appearing white in visible light. Visible and infra-red emissivities of *Dimetrodon* may have been separately adjustable according to the thermal state of the animals" (Bramwell and Fellgett 1973).

In a similar way, the dorsal plates of *Stegosaurus* (Fig. 71) probably acted as forced convection fins, serving an important thermoregulatory function. Wind-tunnel experiments on finned models, calculations of internal heat conduction, and direct observations on the morphology and internal structure of stegosaur plates support this hypothesis, demonstrating the comparative effectiveness of the plates as heat dissipaters, controllable through the rate of input blood flow, temperature, and body orientation with respect to wind (Farlow et al. 1976). The dermal plates of *Stegosaurus*, like other bizarre structures which have, until recently, defied satisfactory explanation, may well also have been concerned with intraspecific combat and display behaviour. Such structures include crests, spines, frills, and dorsal plates (Sect. 9.3; see review by Hopson 1977.)

7.6 Tachymetabolism

There is a fundamental difference between the metabolic rates of homeothermic or tachymetabolic (= fast rate of chemical change) animals and those of heterothermal bradymetabolic (= slow rate of chemical change) animals whose body temperatures tend to vary according to that of the environment. In today's mammals and birds the metabolic rate is about four times greater than that in extant reptiles.

The occurrence of blood shunting from superficial to deep tissues as a means of conserving heat is now well established among modern reptiles. Cowles (1958) argued that it originated from the dermal vascularity of Amphibia and must have served in the remote past as a necessary adjunct to their notable dependence on supplemental dermal respiration. Later, it became a mechanism for collecting and dispersing heat in reptiles and, finally, the essential temperature regulator of the endotherms.

There has been considerable speculation regarding the possible capacity of the dinosaurs to produce metabolic heat to some measure in excess of that of modern reptiles. Benedict (1932) showed that the rate of metabolism in reptiles decreases with increase in body size, while Schatz (1957) claimed that the dinosaurs had extremely slow rates of metabolism. Other workers, however, have maintained that the dinosaurs were not only homeothermic, but were also tachymetabolic, and their arguments have been criticised severely. There are two factors which strongly suggest that the dinosaurs may have differed physiologically from modern reptiles. First, they were the dominant terrestrial vertebrates for over 100 million years, in which respect they have been equalled only by the mammals. Second, they disappeared, apparently quite suddenly, about 65 mya, whereas crocodiles, turtles, lizards and snakes underwent no major change at that time and have even shown a modest amount of adaptive radiation subsequently. Several authors have postulated that the evolution of endothermy could have been a critical factor in the extinction of the Mesozoic archosaurs (Chap. 12).

In 1968, Bakker claimed that dinosaurs were fast, agile, energetic creatures that lived at a high physiological level reached elsewhere among land vertebrates only by the later, advanced mammals (Sect. 7.4). Three years later, he specifically proposed that they might have been homeothermic, and that the production of heat was endothermal (Bakker 1971a). Then he wrote an article in which he summarised all the arguments for endothermy or tachymetabolism in dinosaurs, classifying them broadly as: 'gross anatomical', 'cellular', and 'community structure'. In 1972, he introduced a fourth category of argument, based on palaeolatitudinal zonation (Bakker 1972). Bakker's arguments have been criticised by Feduccia (1973), Bennett and Dalzell (1973), Thulborn (1973, 1975), and Feduccia (1974), among others. They were defended by Ostrom (1974b), Dodson (1974), by Bakker himself (1974), and later by Wilford (1986). Defending arguments were expanded into a controversial book by Desmond (1975): Bakker's own book on the subject appeared rather later (Bakker 1987). A comprehensive and unbiased review of the dispute was given by Charig (1976) who pointed out that although he, himself, was not opposed to the concept of endothermy (which he considered to be neither proved nor disproved), he was opposed to the use of fallacious argument, with which many of the papers cited above would, he claimed, appear to have been liberally endowed!

7.6.1 Anatomical Evidence

Bakker (1971a, 1972) proposed that, since the dinosaurs had a fully erect gait – indicated by the anatomy of their joints and their fossilised footprints – their metabolic rates and levels of activity must greatly have exceeded those of modern reptiles. This interpretation was challenged by Bennett and Dalzell (1973) and Feduccia (1974) on the grounds that no logical connection between fully erect posture and endothermy has been demonstrated. Metabolic data based on studies of modern lizards fail to reveal any correlation between limb length and metabolic rate: at equal body temperature, standard metabolic rate is solely a function of body size. "Dinosaurs may well have been homeothermic, but Bakker fails to construct a convincing case. This failure rests on an inadequate analysis of the biomechanics of fossil and contemporary vertebrates and on the inability to establish a physiological connection between homeothermy, metabolism and posture" (Bennett and Dalzell 1973).

Feduccia (1973) also pointed out that Bakker (1971a) was imprecise in his terminology. "Without defining terms, Bakker (1971a, p. 637) speaks of homeothermic mammals, and concludes (Bakker 1971a, p. 656) with homeothermic dinosaurs. One assumes that he is using homeothermy synonymously with endothermy. But his terminology is confusing when he states (Bakker 1971a, p. 646) 'Endothermy and thermal homeostasis require a larger energy budget than ectothermy and heterothermy, and therefore homeothermy requires more continuous food-getting activity...' Ostrom's statement (1969, p. 369) that dinosaurs, etc., '...were homeothermic and perhaps endothermic...' is also somewhat confusing, but one is generally led to believe that both authors consider the dinosaurs to be very close to birds and mammals in their abilities at temperature regulation" (Feduccia 1973).

The major premise of the arguments of Ostrom (1969) and Bakker (1971a) relied upon the fact that there are no living terrestrial vertebrates with erect posture that are ectothermic. From this, both authors concluded that all erect terrestrial vertebrates must be endothermic but, as Feduccia (1973) replied, the converse does not hold.

Erect posture, vertebral excavations, along with a myriad of other more specific adaptations were probably all part of a complex structural system to support the tremendous weight of these giants. Furthermore, Heath (1968) who also postulated that the dinosaurs might have been endothermal, noted too that higher vertebrates and insects probably developed thermoregulation as a response to rapid fluctuations of temperature in terrestrial environments.

To such criticisms Ostrom (1974b) replied: "Of course an endotherm may be adapted for a sprawling existence with non-erect posture. Endothermy does not impose upright carriage, but ectothermy may be a major constraint that makes erect posture and locomotion impossible." Whereas Dodson (1974) wrote: "It is commonly accepted that the rigors of the climate selected for endothermy among therapsids; it is equally probable that the same selective agents affected a similar state among advanced thecodonts. The fossil record strongly supports the interpretation that dinosaurs attained physiological refinement first, as expressed in erect posture and bipedality, and that mammals avoided competition and extinction by maintaining small body size and inconspicuous habits through the Mesozoic. The remainder of the Era is not to be thought of as a benign hot-house. We owe the preservation of North American Jurassic and Cretaceous dinosaurs to orogenies that produced widespread tectonic deltas, and it may be assumed that these orogenies affected climate as did their counterparts in the Triassic."

7.6.2 Bone Histology

While the quality of dinosaur muscle (in which most thermogenic activity must have taken place) cannot be assessed, one subtle aspect of their anatomy which strengthens the comparison with birds and mammals is the finding, now well established, that the histology of dinosaur bone is characterised by the extensive development of dense secondary Haversian canal systems (see refs. in Dodson 1974; de Ricqlès 1974). In contrast, the bones of living ectotherms are relatively poorly organised, having few Haversian canals and very little reorganisation of spongy tissues. Moreover, in strongly seasonal climates, where drought or winter cold forces ectotherms to become dormant, growth rings appear in the outer layers of compact bone, much like the rings in the wood of trees in similar environments. Such growth rings are hardly ever seen in the bones of tachymetabolic birds and mammals, nor do they appear in those of dinosaurs. Comparative studies of the histology of reptilian bones make possible a functional interpretation of their structure (Fig. 73). This suggests that several orders of fossil reptiles, including therapsids, dinosaurs, and pterosaurs, were tachymetabolic (de Ricqlès 1974, 1976, 1980). According to Bouvier (1977), however, comparative histological data show that secondary Haversian bone is found among both ectotherms and endotherms. Not all endotherms possess extensive Haversian systems nor do all ectotherms lack them. Moreover, the function of these may partly be to increase the mechanical strength of bone. Consequently, Haversian bone should not be utilised as an indicator of thermoregulatory strategy as Bakker has attempted to do. On the other hand, if de Ricqlès (1976, 1980) was correct, his data do strongly support the suggestion that not only dinosaurs but also therapsids (as well as ptero-

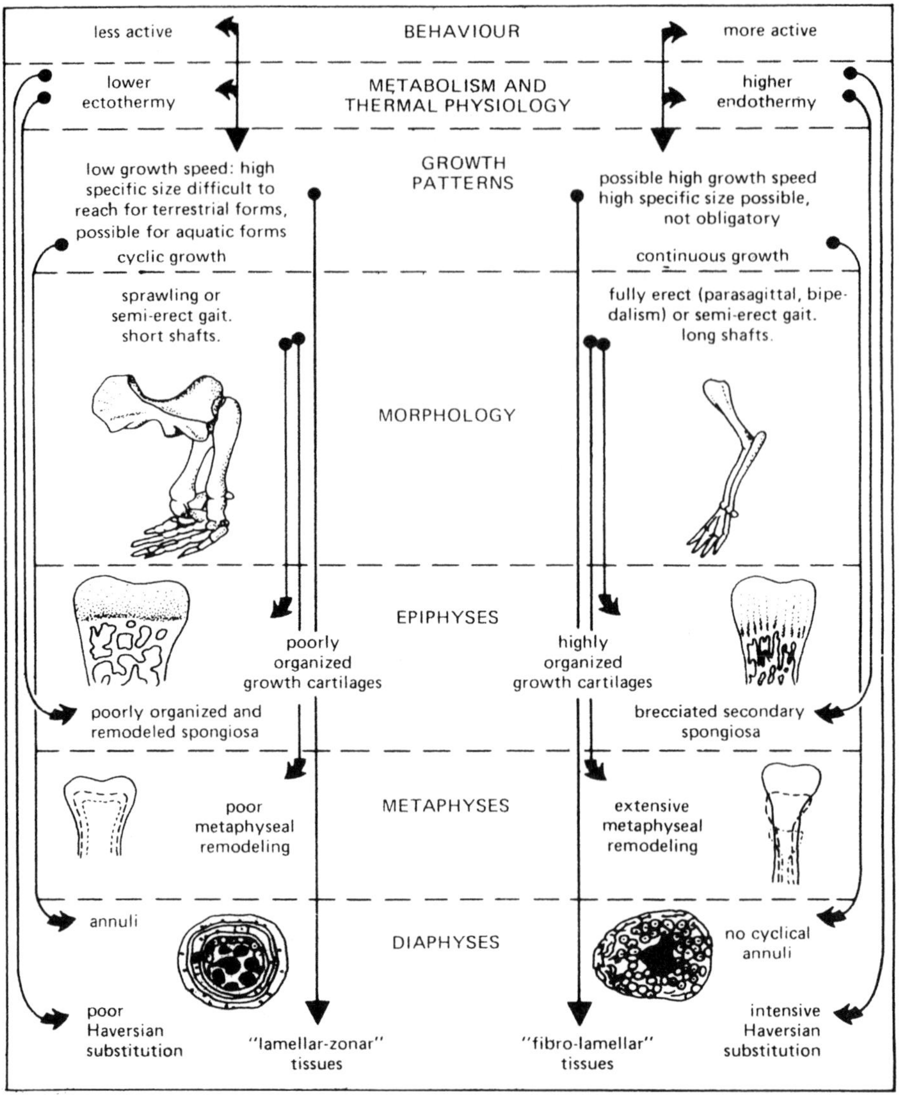

Fig. 73. An attempt to visualise gross functional relationships between bone histology, morphology and thermal physiology among terrestrial tetrapod vertebrates. Typical primitive ectothermic pattern is shown on the *left*, typical advanced endothermic pattern on the *right*. Many groups of tetrapods may of course show various intermediate conditions, for one or several factors involved, and hence should be placed at or near the middle. (Cloudsley-Thompson 1978 after de Ricqlès 1974)

saurs) were endotherms with tachymetabolic thermoregulation (see list of papers by de Ricqlès in Thomas and Olson 1980; Reid 1984, 1997).

7.6.3 Energy Flow and Predator: Prey Ratios

Another indication of thermogenesis is afforded by the ratio of predatory animals to their prey. Thus, by extrapolating from the food consumption of birds and mammals, it is possible to obtain crude estimates of the theoretical ingestion rate of 'tachymetabolic dinosaurs' and similar extrapolation from the ingestion rates of reptiles and amphibians provides theoretical estimates of the ingestion rate of 'bradymetabolic dinosaurs'. By deriving an empirical equation relating the ratio of predatory forms to prey, it should be possible to determine whether the predators were ectothermic or endothermic. This ratio is constant, regardless of the body size of the animals in the system, in consequence of the effects of scaling in predator:prey energy flow; but it differs very considerably according to whether the predator is bradymetabolic or tachymetabolic. Bakker (1975) expressed this argument as follows: "The energy budget of an endothermic population is an order of magnitude larger than that of an ectothermic population of the same size and adult weight, but the productivity...is about the same... A given prey population, either ectotherms or endotherms, can support an order of magnitude greater biomass of ectothermic predators than of endothermic predators, because of the endotherms' higher energy needs. A herd of zebras produces about a quarter or a third of its weight in carcasses per year, but a population of mice may yield up to six times its weight because of its rapid turnover, reflected in short life span and high metabolism per unit weight. Now, the energy budget per unit of predator standing crop also decreases with increasing weight: lions require more than ten times their own weight in meat per year, whereas shrews need 100 times their weight. These two bioenergetic scaling factors cancel each other, so that if the adult size of the predator is roughly the same as that of the prey (and in land vertebrates it usually is), the maximum ratio of predator standing crop to prey standing crop in a steady-state community is a constant independent of the adult body size in the predator-prey system."

Some fossil deposits yield hundreds or thousands of individuals representing a single community. Their live weight or biomass can be calculated from reconstruction of complete skeletons, and the total predator:prey ratios are then easily worked out. Such an analysis of energy flow strongly indicates that the energy budgets of the dinosaurs were like those of large mammals, not elephant-sized lizards (Bakker 1972).

Bakker (1972) based some of his calculations on the estimate that a Komodo dragon (*Varanus komodoensis*) takes large prey about once a month, together with the assumption that the weight of the prey is roughly half that of the lizard. From this he deduced that the Komodo dragon consumes its own weight in prey every 60 days. There is evidence, however, as Thulborn (1973) pointed out, that the animal can kill prey two or three times as heavy as itself, and may ingest its own weight in prey every 10–15 days, whereas the tiger consumes its own body weight in food approximately every 24 days.

Another objection that may be brought against Bakker's argument is that palaeontologists, when collecting fossils, invariably go for the rarer predators rather than for herbivores – which are far less interesting. The bias results in an apparent exaggeration in the biomass of predators. This, however, would slant the ratio in the direction of bradymetabolism, while the enormous size of the herbivores would be bound to distort the ratio even if the vagaries of fossilisation are ignored (Cloudsley-Thompson 1978). Other objections raised by Charig (1976) included the fact that no allowance can be made for differing life spans, that some prey species were more palatable than others, robust skeletons are more likely to be preserved than fragile ones, and species that lived in the lowlands would have been more likely to be preserved than upland species.

Bakker (1972) suggested that the dinosaurs were built for sustained locomotion at moderate speeds and inferred from this that they were endothermic, but Thulborn (1973) considered that they were not built for sustained speed, but rather attained their maximum speed in short bursts. The cheetah can attain a maximum speed of 97 km h^{-1} for about 15–20 s, but soon slows down if not successful in catching its prey. Lizards, although cold-blooded, are also capable of fast running for short periods and a similar mechanism could have permitted sophisticated cheetah-like predation in communities of ectothermic dinosaurs.

In order to test Bakker's (1972) hypothesis, Farlow (1976) investigated the trophic dynamics of a community of large Upper Cretaceous dinosaurs in the Oldman Formation – sediments that were deposited along the margin of a great inland sea that once covered much of the western interior of North America. The environment of deposition appears to have been tracts of fluvial marshes that separated islands of higher, drier ground. The climate was warm-temperate, and Farlow suggested that the upland plant communities were probably park-like in aspect. The large dinosaurs of this community comprised animals that were between a hippopotamus and a large African elephant in adult weight. Calculations suggest that the annual secondary production of tachymetabolic herbivorous dinosaurs would have been insufficient to meet the food requirements of a bradymetabolic carnivorous dinosaur population as large as is preserved in the Oldman Formation. However, ectothermic dinosaurs would have been easily able to make energetic ends meet. Unfortunately the situation is complicated by the possibility that carnivores are overrepresented in collections from Oldman. Because of this, Farlow (1976) was unable to decide between ectothermy or endothermy, and the question still remains undecided. Bakker's idea, though sound in theory, should not be applied in practice (Charig 1976)!

7.6.4
Palaeolatitudinal Distribution

One of the analytical tools of palaeobioenergetics is latitudinal zonation. Palaeomagnetic data makes it possible to reconstruct, to within about 5° latitude, the ancient positions of the continents before they drifted to their present positions. According to Bakker (1975), large reptiles found fossilised in sediments laid down under cold, or seasonally cold, conditions, could not have endured those conditions unless they were endothermic. Although he applied this argument only to mammal-like reptiles, it presumably carries the corollary implication that endothermy evolved as an adaptation to low temperatures in dinosaurs also. In fact, the argument cannot be applied to dinosaurs, as Charig (1976) pointed out, and Bakker's conclusions regarding palaeoclimates were contrary to those of many other workers in the field. If the dinosaurs died of cold, as Bakker (1975) also suggested, this might be taken to imply that they were ectothermic, since it has been argued that endothermy evolved in response to the colder periods of a seasonal climate (Dodson 1974).

Relying primarily on the work of John Ostrom, Robert Bakker, and Armand de Ricqlès on dinosaur ecology, energetics, posture, gait and bone histology, Desmond (1975) presented an extreme view of the case that the dinosaurs were active terrestrial creatures with mammal-like physiology. In his book, he linked endothermy with high intelligence and this, in turn, with complex behaviour. He also asserted that if an adult reptile remains near the young, its limited intelligence cannot overcome the temptation to eat them. However, as R.J. Wassersug pointed out in a review of Desmond's book, "Adult crocodiles co-operate in feeding activities. Parents help their hatchlings escape the egg and nest; a mother collects hatchlings and transports them to the water in her mouth. Complex behaviour of the type posited for dinosaurs evidently does not require endothermy in the archosaurs." Moreover, Desmond (1975) exaggerated the view of de Ricqlès (1974), who wrote, "If the origins of perfect 'warm-bloodedness' (endo- and homeothermy) are looked for among the primitive representatives of lineages of warm-blooded modern vertebrates, one cannot ask for a sudden appearance among them of all the associated features that one can find among living, modern warm-blooded animals." In de Ricqlès' opinion, the larger dinosaurs had a peculiar physiology by any standard, one that can hardly be regarded as typically reptilian, but better understood as something of its own.

Many authors have argued that the dinosaurs were endothermic (Bakker 1968, 1971a, 1972, 1974, 1987; Desmond 1975; Dodson 1976; Heath 1968; Ostrom 1969, 1974b; de Ricqlès 1974, 1976), while others have disputed their arguments, and claimed that the dinosaurs were ectothermic (Bennett and Dalzell 1973; Feduccia 1973; Thulborn 1973, 1975; Bennett 1974). Certainly, they could not have been very speedy (Alexander 1985, 1989, 1997a,b; Thulborn 1973) but bulk alone, as we have seen, may be quite sufficient to retard the dissipation of environmentally acquired heat (Colbert et al. 1946; Spolita et al. 1973). Seymour (1976) argued that the sustained high blood pressure inferred from the large vertical distance between the heart and the head in some dinosaurs (Sect. 7.3) supports, but does not prove, the proposition that dinosaurs were endothermic. The case for tachymetabolism cannot be regarded as proven, even though the dinosaurs must, almost certainly, have been homeothermic. Bennett and Dawson (1976) concluded that, if the dinosaurs did possess an elevated metabolism, it may have evolved in response to factors involved in activity rather than in thermoregulation.

Some years later, Bennett and Ruben (1986) reviewed the considerable amount of evidence for endothermy in therapsids. This included the nasal turbinal complex (the nasal cavities approached the condition found in modern mammals), the large number of traits shared by therapsids and monotremes, and the histology of their bones. These authors concluded that at least the more advanced of the mammal-like reptiles were tachymetabolic, the adaptive advantage of which had been discussed by Kemp (1982; Sect. 7.5.6).

7.7 Bradymetabolic Thermoregulation

In comparison with the reptiles of the Mesozoic Era, modern reptiles are mostly small and bradymetabolic. Their large surface-to-volume ratio would render tachymetabolism uneconomical. Furthermore, they are able to evade inclement weather by aestivating in summer or hibernating in sheltered retreats. Even so, in addition to behavioural temperature regulation, they are known to utilise a number of physiological thermoregulatory processes. These include the emergency cooling of the body of tortoises through salivation and urination, and the flexing of the body muscles of pythons – which increases metabolic heat production when brooding their eggs (see Cloudsley-Thompson 1999). There seems little doubt that the dinosaurs and other large Mesozoic reptiles must have possessed some physiological thermoregulatory processes. In view of their large size and small surface-to-volume ratio, these would have been much more effective than those of small modern reptiles. This argument applies equally, whether they were tachymetabolic or bradymetabolic. They were still probably homeothermic, maintaining fairly constant body temperatures.

In warm climates, giant reptiles would have some of the characteristics of existing homeotherms, even if they had a low metabolic rate. The behaviour of these reptiles could therefore have been relatively independent of diurnal changes in radiation and air temperature – so long as the average values of these parameters resulted in a body temperature that was within the limits of tolerance. Large size, and the consequent high capacity for storing heat, would have dampened out environmental changes over periods of several days and maintained body temperatures at relatively constant levels. Changes in the thickness of the subdermal fat layer would have been of minor importance in contributing to the overall stability of the body temperature, according to the calculations of Spolita et al. (1973) who concluded that gigantism was, no doubt, a very useful strategy for large reptiles, providing a constant, equable internal temperature in the absence of a high metabolic rate when living in a stable warm climate. Dennis Claussen has recently calculated that if an animal as large as *Argentinosaurus* (Sect. 10.4) were tachymetabolic, to pump blood up to its head would have required a heart with walls some 0.5 m thick and weighing about seven tonnes. If, however, it was bradymetabolic, a much smaller heart would have sufficed (cited in Natl Geogr Mag March 2003: 2–33).

7.8 Hair and Feathers

It is now generally assumed that the pterosaurs were hairy (Sect. 6.3). They were almost certainly tachymetabolic also, according to Fleming and Lips (1991) (Sect. 6.7.1). In recent years, several fossils of non-avian therapod dinosaurs with well-preserved integumentary coverings have been discovered. Apart from numerous specimens of *Caudipteryx,* which possessed true feathers (Ji et al. 1998), however, fossils of only one or two other genera of feathered dinosaurs have so far been discovered. One of these, belonging to the family Dromaeosauridae, has been recently discovered in Liaoning Province, China (Ji et al. 2001). Dating back ca.130 my, it was covered from head to tail in a downy fluff of feather-like fronds. This makes it indisputable that a body covering similar to feathers was present in some non-avian dinosaurs. The dromaeosaurs were advanced therapods (Chap. 11) and were probably more like birds in appearance than giant lizards. The patterns of feathers on the arms and tail were similar to those of *Caudipteryx zoui.* Much of the controversy over the origin and early evolution of birds (Sect. 11.5) centres on whether or not they were derived from caelurosaurian therapod dinosaurs. These animals from the Upper Jurassic and Lower Cretaceous probably could not fly. Phylogenetic analysis showed them to be even more primitive than *Archaeopteryx,* the earliest known bird. They represent stages in the evolution of birds from feathered, ground-living bipedal dinosaurs and show that the origin of feathers was unrelated to the origin of flight.

Although most authors accept that crocodiles and birds are sister-groups (Benton 2004), Brian Gardiner (1982, 1993) produced a radical alternative hypothesis based on a character analysis of living tetrapods. He concluded that mammals and birds are sister-groups (Supercohort Haematothermia), sharing numerous specialisations with crocodiles and, before them, Chelonia (Sect. 11.5). This interesting speculation lies somewhat beyond the scope of the present volume, but it does support the hypothesis that the primary function of both feathers and hair was initially thermoregulatory.

7.9 Conclusions

The adaptive value of a permanent and high constant body temperature depends upon the action of enzymes. A complex, multi-enzyme system must evolve and function, because each individual enzyme system is constant and therefore predictable. Highly complex biological organisation depends upon multi-enzyme systems. No single characteristic that responded to numerous condition could evolve unless it were accompanied by all the other characteristics that were necessary. There must therefore have been a gradual evolution of all such systems, and many of the reptiles in the fossil record provide evidence of poor homeostasis. The disadvantages of tachymetabolism included, among others, greater food requirements, and the necessity for more complex digestive systems, the evolution of methods for cooling as well as for heating the body and enhanced locomotor abilities (Kemp 1982).

The large size of most dinosaurs obscures the distinction between tachymetabolic and bradymetabolic thermoregulation, for which most criteria are size-dependent. Likewise, predator:prey ratios do not distinguish between tachymetabolism and bradymetabolism because these converge as the sizes of the animals concerned increase. Bone histology shows only that the dinosaurs were different from living reptiles, not that they were similar to mammals. Water conservation and locomotor refinements were more important than thermal strategy to dinosaur success. Finally, size distinguishes dinosaur evolution from that of extant endotherms. Large size led to reduced rates of heat-transfer between the reptile and its environment, resulting in inertial homeothermy. If necessary, smaller dinosaurs could hibernate or aestivate; larger dinosaurs would not need to do either (Hotton 1980). (Indeed, hibernation and aestivation would probably not have been necessary, even in the case of the smaller dinosaurs.) Hotton (1980) also pointed out that all really large dinosaurs would have had to have been nomadic in order to be able to gather enough food. As seasonal climates became widespread and persistent during the Mesozoic Era (as indicated by growth rings in fossil timber) they would have influenced the adaptive evolution of the dinosaurs and engendered long range seasonal migrations. Evidence for this is, he claimed, readily obtained from elements present in Upper Cretaceous palaeogeography. Migration and random drift over large distances depended on size, locomotor ability, and also on continuous activity for the great distances that would have had to have been travelled.

It would seem that the larger dinosaurs must have been homeothermic and to some extent tachymetabolic, but they could not possibly have been active to the degree indicated by Bakker (1971a, 1972, 1987) and Desmond (1975). The controversy has been reviewed very clearly by Fastovsky and Weishampel (1996) and Benton (2004). Farlow (1997), Reid (1997), in addition to various authors in Currie and Padian (1997) and in Ruben et al. (1997), are recommended for further reading.

8 Therapsids, Anapsids and Early Diapsids

8.1 Introduction

During the Permian period, the dominant reptiles were pelycosaurs (Sect. 2.5) but, in the Triassic, these were replaced by therapsids or mammal-like reptiles (Sect. 2.6) from which the mammals evolved. During the same period, the early archosaurs of the Permian gave rise to thecodontians which were ancestral to the crocodilians (Sect. 4.6.3), the pterosaurs (Chap. 6) and the dinosaurs of the Mesozoic Era. The reptiles in nearly every one of these and other evolutionary sequences show an increase in size as time passed. Almost never, it seems, did big vertebrates evolve into smaller forms. On the contrary, they appear nearly always to have grown larger and larger until eventually they became extinct for one reason or another!

8.2 Therapsids and the Origin of Mammals

The synapsid order Therapsida (Sect. 2.5) may have diverged from the pelycosaurs during the Lower Permian, but the earliest fossils to be found were in the Upper Permian of Gondwanaland. Therapsids diversified greatly in the Triassic when they dominated the terrestrial vertebrate forms of the world; but only one suborder, Cynodontia, persisted into the Lower Jurassic. The preceding Dicynodontia had been the most successful herbivorous therapsids of the Late Permian and Triassic – a span of almost 50 my. These probably had a variety of masticatory movements, cropping with the horny beaks at the front of the mouth and grinding the food with the horn-covered palatal surfaces further back (King 1996). Triassic genera included *Lystrosaurus* (Fig. 74) from Antarctica which, at one time, was thought to have wallowed in shallow water like a hippopotamus, feeding on aquatic weeds. It is now believed to have browsed on more resistant plants and might even have been a burrower. This could have been an adaptation to an increasingly semi-arid climate (Palmer 1999). *Lystrosaurus* was one of the forms on which Alfred Wegener in 1912 based his evidence in favour of continental drift (Sect. 2.2.2).

Another Triassic dicynodont was *Kannemeyeria* (Fig. 74) from South Africa, India, and Argentina – further evidence for the existence of Gondwanaland. It had a massive head with unusually large openings for the eyes, nostrils and jaw

Fig. 74. Dicynodonts. *Above Lystrosaurus* (Lower Triassic; length ca. 1 m). *Below Kannemeyeria* (Middle Triassic; length ca. 3 m). (After Palmer 1999)

muscles. It evidently tore up roots, stripped leaves from the vegetation with its horny break and ground them up with its toothless jaws (Palmer 1999). The dicynodonts had large abdomens which would have housed capacious gastrointestinal systems – as might be expected in animals that consumed large quantities of vegetable material, to be retained in the alimentary canal for quite a long time. This suggests that they did not digest their food as quickly as would have been the case if they had been tachymetabolic like the later therapsids.

The dicynodonts were the first vertebrate herbivores to become really numerous. With their appearance, the primary production of vegetable matter could, at last, have been harvested directly. As a result, a completely new ecological system was set up (see King 1996).

The cynodonts were by far the most successful of the Therapsida and showed many advanced mammalian features. Even *Cynognathus* (Fig. 75) from the Lower Triassic of South Africa and Argentina probably had a coat of fur, and later forms almost certainly did. *Cynognathus*, one of the largest of the cynodonts, was a ferocious predator, strongly built and with its hind legs almost directly beneath its body. Other Triassic cynodonts included *Thrinaxodon* (Fig. 76a) from South Africa and Antarctica and *Massetognathus* (Fig. 76b) from Middle Triassic strata of Argentina. *Oligokyphus* (Fig. 76c) from England was a representative of the tritylodonts, the only group to persist into the Jurassic.

Thrinaxodon (Fig. 76a) was much more mammal-like than its earlier relations. It was a small, but sturdily built carnivore. Its body was long and clearly divided into thoracic and lumbar regions, the thoracic vertebrae forming a rib

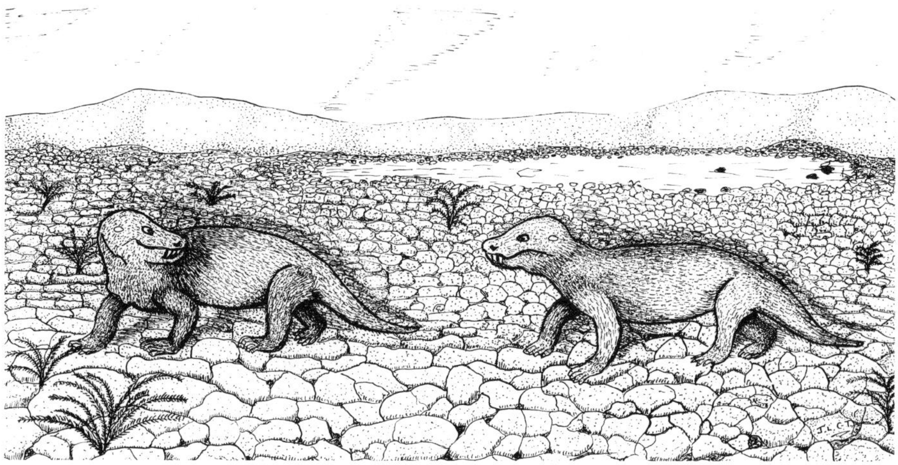

Fig. 75. *Cynognathus* (Cynodontia; Lower Triassic; length ca. 3 m). (Cloudsley-Thompson 1999)

cage, a feature not previously found among vertebrates. *Thrinaxodon* was evidently quite speedy, thanks to its erect posture and strong hind legs. The rib cage was probably closed by a diaphragm that would have filled and emptied the lungs very efficiently, while a secondary bony palate separated the breathing passage from the mouth, thus enabling the animal to breathe whilst chewing its food into smaller pieces for easy digestion. These developments give a strong indication that *Thrinaxodon* was tachymetabolic. *Massetognathus* (Fig. 76b) and *Oligokyphus* (Fig. 76c) were both herbivorous – as indicated by their respective dentitions, despite the weasel-like appearance of the latter. *Oligokyphus* had evolved a fully upright four-legged posture – the only therapsid not to have sprawling forelimbs. Its dentition, too, was significantly different from that of other cynodonts. There were no canines and the front incisors were greatly en-

Fig. 76a–d. Later therapsids. **a** *Thrinaxodon* (Cynodontia; Lower Triassic; length ca. 50 cm), **b** *Massetognathus* (Cynodontia; Middle Triassic; length ca. 50 cm), **c** *Oligokyphus* (Cynodontia; Lower Jurassic; length ca. 50 cm), **d** *Ericiolacerta* (Therocephalia; Lower Triassic; length ca. 20 cm). (After Palmer 1999)

larged, resembling those of rodents. It was so mammal-like in appearance that, for many years, it was believed to have been a mammal! (Palmer 1999).

The origin of mammals presents no real scientific problem. The immediate ancestors of the class were undoubtedly cynodonts. Closely related to them, was another advanced therapsid suborder, Therocephalia. Its members ranged in size from small insectivores to large carnivores and, in the Early Triassic, some therocephalians such as *Bauria*, from South Africa, became successful herbivores (Benton 2004). *Ericiolacerta* (Fig. 76d) was an active insectivore with small teeth and relatively long limbs. The abundant vegetation of the Lower Triassic, consisting of ferns, conifers, and cycads, not only supported large numbers of dicynodonts but also provided abundant arthropods on which preyed lizard-like insectivores such as *Ericiolacerta.* The first mammals are believed to have been small forms such as *Adelobasileus* and *Sinoconodon* of the Upper Triassic, which escaped the attentions of predatory reptiles by their nocturnal habits.

Towards the end of the Triassic period, the Therapsida virtually disappeared and, within a comparatively brief space of time, the dinosaurs radiated until they had occupied nearly all terrestrial niches. They then ruled the land for the next 120 my. Many palaeontologists have assumed that the decline of the mammal-like reptiles was gradual and that dinosaurs and their ancestors, thecodontians, rose to dominance during that time. Benton (1983) reviewed the subject and presented evidence suggesting that the takeover had not been a result of competition. Alan Charig (1984), however, claimed that the Triassic faunal replacement was certainly competitive, at least in the sense that of two sympatric lineages occupying the same broad adaptive zone, and subjected to the same environmental pressures, one reached total doninance, while the other waned almost to the point of extinction.

8.3 Anapsids

The subclass Anapsida (Sect. 2.4) is represented by a number of early reptilian taxa including the mesosaurs (e.g. *Mesosaurus*, Fig. 7), the procolophonids and the Testudines. The family Procolophonidae ranged throughout the world from the Upper Permian to the Upper Triassic. Its early members were small and agile insectivores, with numerous peg-like teeth. *Procolophon* (Fig. 77) from the Lower Triassic of South Africa and Antarctica was probably herbivorous or insectivorous. Later members, from the Middle Triassic onwards, were usually larger, herbivorous, and with broad cheek teeth suitable for grinding rough plant material. The spikes around its head were probably defensive. Apart from the Testudines, aquatic members of which were discussed in Chapter 4, few anapsids survived into the Mesozoic Era.

The Triassic *Proganochelys* and *Triassochelys* (Fig. 18) were remarkably similar to modern terrestrial terrapins except that they were unable to retract their heads and legs into the shell as completely as modern Testudines can. They

Fig. 77. Procolophonids. *Left Procolophon* (Lower Triassic; length ca. 40 cm). *Right Hypsognathus* (Upper Triassic; length ca. 30 cm)

probably spent their lives grazing on low vegetation, and buried their eggs in the ground. Although the horn of their shells did not ever become fossilised, its presence is indicated by marks on the bones. So their defence against predators must have been exactly like that of extant forms.

The most successful group of chelonians – the suborder Cryptodira – evolved, along with the Pleurodira, during the Jurassic period. The Pleurodira, it may be remembered, retract their heads by flexing the neck sideways. Most of them were, as they still are, aquatic or amphibious. Many of the cryptodires retract their head into the shell by lowering the neck and pulling it directly backward. By the end of the Jurassic, the order had greatly diversified and had replaced the pleurodires in the seas, rivers, and lakes throughout the world. Cryptodires are represented in the world today by land tortoises, aquatic and amphibious terrapins, and marine turtles. Tortoises are vegetarian, while amphibious terrapins and marine turtles are more often omnivores or predatory carnivores. Food intake can be influenced by its availability. Moreover, many taxa of Testudines tend to be carnivorous when young, but switch to becoming largely vegetarian as they get older. Most probably, the same would have been true of Mesozoic forms (Sect. 4.3.2).

8.4 Early Diapsids

8.4.1 Lepidosauria

The Diapsida, with two temporal fenestrae, includes the lizards and snakes, crocodilians, pterosaurs, dinosaurs and birds. The crocodilians and pterosaurs have already been discussed. Here, we consider the ancestral lizards and snakes

Fig. 78. *Planocephalosaurus* (Rhynchocephalia; Upper Triassic; length ca. 20 cm). (Based on Palmer 1999)

(Squamata) and Rhynchocephalia or Sphenodontida (Table 2), which are included with them in the infraclass Lepidosauria. The lepidosaurs first appeared in the Late Triassic – over 200 mya and diversified into numerous families during the Jurassic period. *Planocephalosaurus* (Fig. 78) was one of the earliest of the sphenodonts to evolve. Its skeleton, from the Upper Triassic of Europe, was almost identical to that of the modern tuataras (*Sphenodon punctatus* and *S. guntheri*) of New Zealand.

The teeth of *Planocephalosaurus* were fused to the jaws rather than being attached to grooves in the jaw bones as they are in advanced lizards. Although smaller than *Sphenodon* (length ca. 75 cm), it could have delivered a powerful bite. It must have preyed on insects, scorpions, worms, snails and even small lizards. The long, slender bones of *Planocephalosaurus* were very much like those of modern lizards but its snout was perhaps blunter than that of most. Later, sphenodontians included some bizarre Late Jurassic and Early Cretaceous forms from North America, with broad grinding teeth (Benton 2004). Pleurosaurs were specialised sphenodonts that evolved on land but returned to the water during Early Jurassic times. The limbs were reduced and the body greatly elongated. In some species, there were up to 57 vertebrae which is about twice the number typical of the order. *Pleurosaurus* (Fig. 26) is an example. It swam rapidly, like a snake, with sinuous movements, its limbs playing no part. *Pleurosaurus* must have been an amphibious predator, coming ashore to mate and lay its eggs (Sect. 4.5).

8.4.2 Squamata

Lizards (Lacertilia) and snakes (Serpentes) are undoubtedly the most numerous and successful reptiles in the world today. Lizards are a much more ancient group than the snakes which evolved from them. The earliest known squamates

Fig. 79. Squamates. *Left Ardeosaurus* (Gekkota; Upper Jurassic; length ca. 20 cm). *Right Pachyrachis* (Serpentes; Lower Cretaceous; length ca. 1 m)

were small, insectivorous, lizard-like reptiles that inhabited southern Africa during the Permian period. The five taxa known today – Iguania, Gekkota, Amphisbaenia, Scincomorpha, and Anguimorpha – must have arisen during the Jurassic, as did the Serpentes, although the oldest fossils known are mostly Cretaceous.

Extinct Mesozoic squamate taxa included the aerial Kuehneosauridae (Sect. 6.2) and other gliding lizards discussed in Chapter 6, as well as the Mosasauria, thalattosaurs, dolichosaurs, and other aquatic taxa (Sect. 4.5). The Gekkota (geckos) was one of the first of the modern groups of lizards to appear. It is exemplified by *Ardeosaurus* (Fig. 79) from Late Jurassic deposits in Germany. *Ardeosaurus* had a flattened head and large eyes, so it was probably a nocturnal predator that fed on insects, arachnids, and smaller lizards. However, whether it possessed adhesive pads on its feet is, of course, not known – they would not have been fossilised. Fragments of other gekkos, as well as of scincomorphs and anguimorphs, have been found in Middle Jurassic deposits of Britain. The Iguania, which includes iguanas, agamids, and chameleons, arose in the Late Cretaceous. The Amphisbaenia is not known from the fossil record of the Mesozoic.

The ancestors of the sixth squamate taxon, the Serpentes or Ophidia, are not known. Fossil snakes, such as *Pachyrachis* (Fig. 79), date back to the Lower Cretaceous. These early forms must have killed their prey by suffocation as do extant boas, pythons, and some Colubridae. They coiled tightly around the ribcage of their victim, tightening the grip every time it exhaled until it had been asphyxiated. Then they swallowed it whole. The Serpentes radiated greatly during the Tertiary when their mammalian prey diversified: poisonous forms did not appear until the Late Eocene. The ecology and behaviour of Mesozoic squamates must have been remarkably similar to that of extant forms (Cloudsley-Thompson 1999) and will not be discussed further here.

8.4.3 Early Archosauromorphs

Therapsid reptiles (Sect. 8.2) suffered badly during the great Permian extinction and, although a few of them became extremely successful and dominant in specific niches during the Triassic, a number of medium-sized carnivorous diapsids, such as *Proterosuchus* (Fig. 80a) and other Lower Triassic archosaurs, took over many ecological niches previously occupied by therapsids. *Proterosuchus* was a slender creature that preyed on therocephalians, dicynodonts and procolophonids. *Chasmatosaurus* (Fig. 80b) from South Africa and Asia, had robust limbs with five digits. These were set at an angle to the body and resulted in a sprawling, lizard-like gait. It may well have spent much of its time in water, preying on fishes. Its numerous sharp and backwardly-curving teeth were well adapted for gripping slippery prey. There were also teeth on the palate, a primitive feature absent from later archosaurs.

Early in the Middle Triassic, a number of archosaur lineages flourished comparatively briefly. One of these contained the proterosuchids, another the fam-

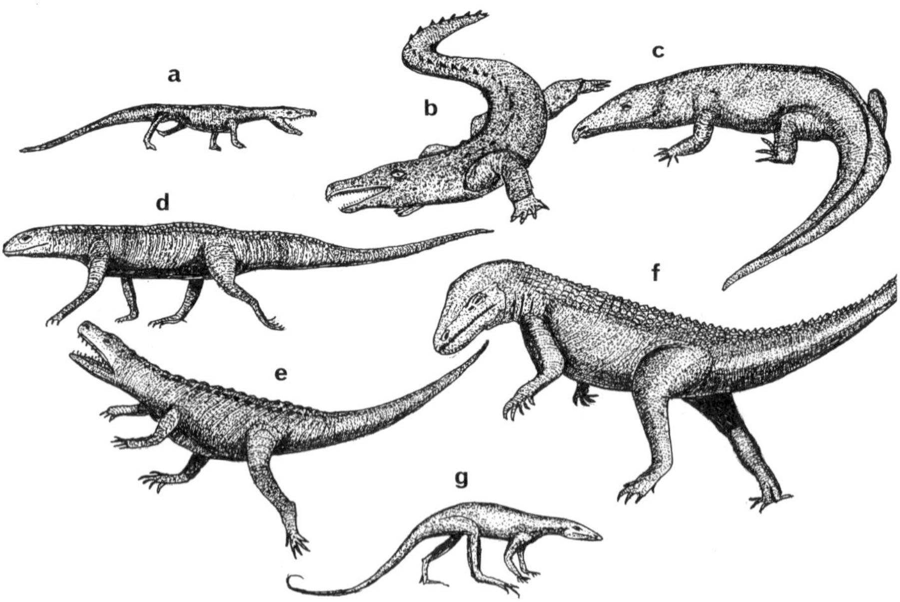

Fig. 80a–g. Early archosauromorph carnivores. **a** *Proterosuchus* (Proterosuchidae; Lower Triassic; length ca. 1.5 m), **b** *Chasmatosaurus* (Proterosuchidae; Lower Triassic; length ca. 2 m; after Charig 1979). **c** *Erythrosuchus* (Erythrosuchidae; Lower Triassic; length ca. 4.5 m ; after Palmer 1999), **d** *Ticenosuchus* (Rauisuchidae; Middle Triassic; length ca. 3 m; after Palmer 1999), **e** *Euparkeria* (Ornithosuchidae; Lower Triassic; length ca. 60 cm; after Palmer 1999), **f** *Ornithosuchus* (Ornithosuchidae; Upper Triassic; length ca. 4 m; after Palmer 1999), **g** *Lagosuchus* (Ornithosuchidae; Middle Triassic; length ca. 30 cm; after Palmer 1999)

ily Erythrosuchidae. The latter comprised a number of top predators including *Vjushkovia*, from Russia, which reached a length of 5 m and fed on dicynodonts and other large herbivores. Another, *Erythrosuchus* (Fig. 80c) from South Africa, was among the most formidable of the Early and Middle Triassic erythrosuchids. It had powerful jaws, studded with sharp, conical teeth, and probably preyed on large herbivores such as dicynodonts and Stagonolepididae (see below).

The amphibious crocodile-like family Phytosauridae has already been considered (Sect. 4.6.2), but a number of other crocodile-like families were terrestrial. One of these, the Rauisuchidae, contained the major carnivorous land animals of the Middle Triassic and was represented by fossils in Europe, Africa, and America. The limbs of these animals were held almost directly beneath the body. This can be seen when *Ticinosuchus* (Fig. 80d) is compared with the phytosaur *Rutiodon* (Fig. 30). The terrestrial ancestors of the crocodilians have already been described and illustrated (Sect. 4.6.3).

Benton (2004) gave a very clear account of the evolution of archosaur ankles and posture. At 3 m in length, *Ticinosuchus* was a medium-sized rauisuchid. In contrast, *Saurosuchus* from the Late Triassic of South America reached a length of up to 7 m and had an erect gait. Fossils of it have been found in association with a rich fauna of dicynodonts, cynodonts, stagonolepidids, small dinosaurs and other reptilians, on all of which it probably preyed. *Saurosuchus* was one of the largest predators to evolve before the appearance of the giant dinosaurs.

In South Africa, *Euparkeria* (Fig. 80e) may have been capable of walking both on all fours and bipedally. An early member of the family Ornithosuchidae, it came at the beginning of the first radiation of the archosaurs. Its hind legs were about a third longer than the forelegs, and its long tail would have balanced its body at the hips when it ran on two legs. This bipedal stance was characteristic of the carnivorous dinosaurs that appeared at the end of the Triassic period. Although small and slim, with light body plates down the centre of the back and tail, *Euparkeria* was a carnivore with long, sharp teeth curving slightly backward and serrated along the edges. *Ornithosuchus* (Fig. 80f) from the Late Triassic of Europe was considerably larger. Although it probably still moved around mainly on all fours, reserving bipedal locomotion for speedy movement when escaping from even bigger predators, it looked very much like a dinosaur. Finally, the diminutive *Marasuchus* (Fig. 83a) was the most dinosaur-like of all the ornithosuchids and unlike *Lagosuchus* (Fig. 80g) may, according to Palmer (1999), actually have been ancestral to the dinosaurs.

In contrast to all other thecodonts the Stagonolepididae or aetosaurs were almost certainly herbivorous. *Stagonolepis* (Fig. 81) from Scotland had a small skull suggesting that food was processed mainly in the alimentary canal rather than in the mouth. Its body was almost entirely encased in armour composed of body scutes, rather like those of crocodiles. The teeth were the same in shape throughout both jaws, and the lack of replacement teeth may imply that replacement was rapid – a necessary response to heavy wear from tough plants such as ferns, horsetails, and the recently evolved cycads. The snout was flat-

Fig. 81. Early archosauromorph herbivores. Above: *Stagonolepis* (Stagonolepididae; Upper Triassic; length ca. 3 m). Below: *Desmatosuchus* (Stagonolepididae; Upper Triassic; length ca. 5 m). (After Palmer 1999)

tened, extremely thin, and pointed. Although the jaws could have snapped shut very suddenly, they would not have produced much force until they had been closed, a capability of more importance to a herbivore than would a more kinetic initial bite as inflicted by carnivores (King 1996). *Stagonolepis* had a small head and no teeth at the front of its foreshortened jaws. The flattening of the pig-like snout would have been a useful adaptation for rooting in the undergrowth.

Desmatosuchus (Fig. 81) from the Late Triassic of North America had particularly heavy body armour (Sect. 9.2.2). Its back, tail and part of the belly were protected by heavy scutes, while pointed spines lined the sides of its back and tail. Larger spines, up to 45 cm in length, projected from the shoulders. Such armour was no doubt necessary to provide protection from the numerous thecodont predators of the time. The deep bodies of the Stagonolepididae were required to accommodate the long intestines needed for digesting vegetable food (Palmer 1999).

Although the archosaurs that we have been discussing were the main group of archosauromorphs to become dominant during the Triassic period, they were not the only ones to appear. The Trilophosauridae was a family of unusual diapsids which had lost the lower temporal opening (Fig. 2) so that their skull was euryapsid in appearance. (Euryapsid reptiles had a single temporal lobe placed high on each side of the skull.) *Trilophosaurus* (Fig. 82) from the Upper Triassic of Texas had a heavily-built skull with flattened three-ridged teeth suitable for shearing tough plant material. Its tail was unusually long. In many parts of the world, the Rhynchosauridae was the dominant family of herbivores in its time (Early to Late Triassic). *Hyperodapedon* (Fig. 82b) was a typical example. Numerous fossils have been obtained from Elgin, in Scotland. Seen from above, the skulls of rhynchosaurs were triangular in shape with the back wider than the front-to-back distance. Space was thus provided for powerful jaw muscles. *Hyperodapedon* had a powerful 'beak' with which to bite

Fig. 82. *Above Trilophosaurus* (Trilophosauridae; Upper Triassic; length ca. 2 m). *Centre Hyperodapedon* (Rhynchosauridae; Upper Triassic; length ca. 1.3 m). *Below Tanystropheus* (Prolacertiformes; Middle Triassic; length ca. 3 m)

seed-ferns and other tough vegetation. Massive claws on the hind feet were used for scratching backwards and uncovering succulent roots and tubes (Benton 1997b).

A fourth group of archosauromorphs, the order Prolacertiformes, first appeared in the Middle Permian and radiated in the Triassic. Most of them looked like large lizards with long necks. *Tanystropheus* (Fig. 82c) from Europe and the Middle East was remarkable in that its neck was more than twice as long as its trunk. Composed of only 9–12 cervical vertebrae, this neck was not very flexible, and its function is a mystery. Juveniles had short necks that grew very rapidly. Wild (1978) has suggested that *Tanystropheus* may have lived in coastal waters, feeding on small fishes it caught by darting its head about very rapidly. Alternatively, it might have lived on the shoreline, dipping its head into the water to catch fish or molluscs that it crushed with its peg-like teeth (Morales 1997). Its competitors in the Triassic seas would have been nothosaurs, placodonts, and ichthyosaurs (Chaps. 4, 5).

Considerably less is known of the ecology and behaviour of the reptilian groups discussed in the present chapter than is known of the Dinosauria. These animals will be dealt with in detail in the three following chapters.

9 The Dinosaurs: Weapons, Display and Reproduction

9.1 The Earliest Dinosaurs

The Dinosauria arose in the Upper Triassic period, some 230 mya, and dominated the terrestrial faunas of the world for the next 165 my. The earliest genera were medium-sized, three-toed bipedal carnivores that grabbed their prey with the forelegs. A sister group is represented by the ornithosuchid or basal dinosauromorph *Marasuchus* (Fig. 83a) (Sect. 8.4.3), a lightly-built flesh-eater that presumably caught small and speedy animals such as cynodonts and procolophonids, as well as feeding, no doubt, on worms, arthropods, and the like

Fig. 83a–f. Early Upper Triassic dinosaurs. **a** *Marasuchus* (Dinosauromorpha; length ca. 1.2 m), **b** *Eoraptor* (Theropoda; length ca. 1 m), **c** *Herrerasaurus* (Herrerasauridae; length ca.11 m), **d** *Staurikosaurus* (Herrerasauridae; length ca. 3 m), **e** *Plateosaurus* (Plateosauridae; length ca. 7 m), **f** *Massospondylus* (Plateosauridae; length ca. 4 m). (Adapted from Lambert 1992 and original)

(Benton 2004). Although *Eoraptor* (Fig. 83b) from Argentina and the Herrerasauridae (Fig. 83c, d) appear to have been true theropods, there is some disagreement as to whether they evolved before or after the split between Ornithischia and Saurischia (Benton 2004). The taxonomy and systematics of dinosaurs have been outlined by Holtz and Brett-Surman (1997).

Large quadrupedal herbivorous dinosaurs had also appeared before the end of the Triassic period. Probably the best known of these is *Plateosaurus* (Figs. 65, 83e), of which numerous skeletons and fragments have been recovered from Germany, France, England and Switzerland. Another member of the Plateosauridae was *Massospondylus* (Fig. 83f), the most common prosauropod of southern Africa. Specimens have also been unearthed in North America. *Massospondylus* was named by Richard Owen in 1854 on the basis of a few large vertebrae. Its small head was carried on a long, flexible neck. Like *Plateosaurus*, it probably walked about on all four legs, rearing up on its hind limbs when necessary to obtain leaves from higher levels of vegetation (Sect. 10.3). *Plateosaurus* had weak, leaf-shaped teeth which, in association with its size and abundance, strongly indicate that it must have been herbivorous. Some controversy has arisen because dagger-like teeth have also been found mingled with the bones of plateosaurs. The explanation is that these carnivorous teeth were almost certainly shed by rauisuchians scavenging on plateosaur carcasses (Lambert 1992; Benton 1997b). The prosauropods such as *Plateosaurus* (Fig. 83e) and *Massospondylus* (Figs. 65, 83f) of the Late Triassic were at one time believed to be ancestral to the true sauropods of the Jurassic. After all, they had only to increase in size and return to a fully quadrupedal mode of locomotion! They are now, however, regarded as a side-branch of the sauropodomorphs that died out at the end of the Triassic period (Moody 1977; Charig 1979; Benton 2004) (Sect. 10.3).

The first of the true dinosaurs was *Eoraptor* (Fig. 83b) from Argentina, a light-weight animal measuring only about 1 m in length. The fossil does not exhibit the specialisation of major dinosaur clades – Ornithiscia, Sauropodomorpha and Theropoda. Nevertheless, its forelimbs were less than half the length of the hind limbs, indicating a bipedal posture, while its functionally tridactyl grasping/raking hands and other features show that it was allied phylogenetically to the theropods. Its discovery supports the hypothesis that the dinosaurs diverged rapidly from a common ancestor while they were small, and that the principal carnivorous and herbivorous lineages were present in the Middle Carnian (Sereno et al. 1993). (The Carnian age was the early part of the Upper Triassic period, 230–225 mya.)

More heavily built were *Herrerasaurus* (Fig. 83e), *Staurikosaurus* (Fig. 83d), and *Pisanosaurus*, all from South America, as well as the North American *Coelophysis*. *Staurikosaurus* and *Herrerasaurus* were members of the same family (Herrerasauridae), and several specimens of the latter genus have been discovered over the years. When these animals first evolved, in every case they comprised only minor components of the faunas to which they belonged (Benton 1990b, 2004).

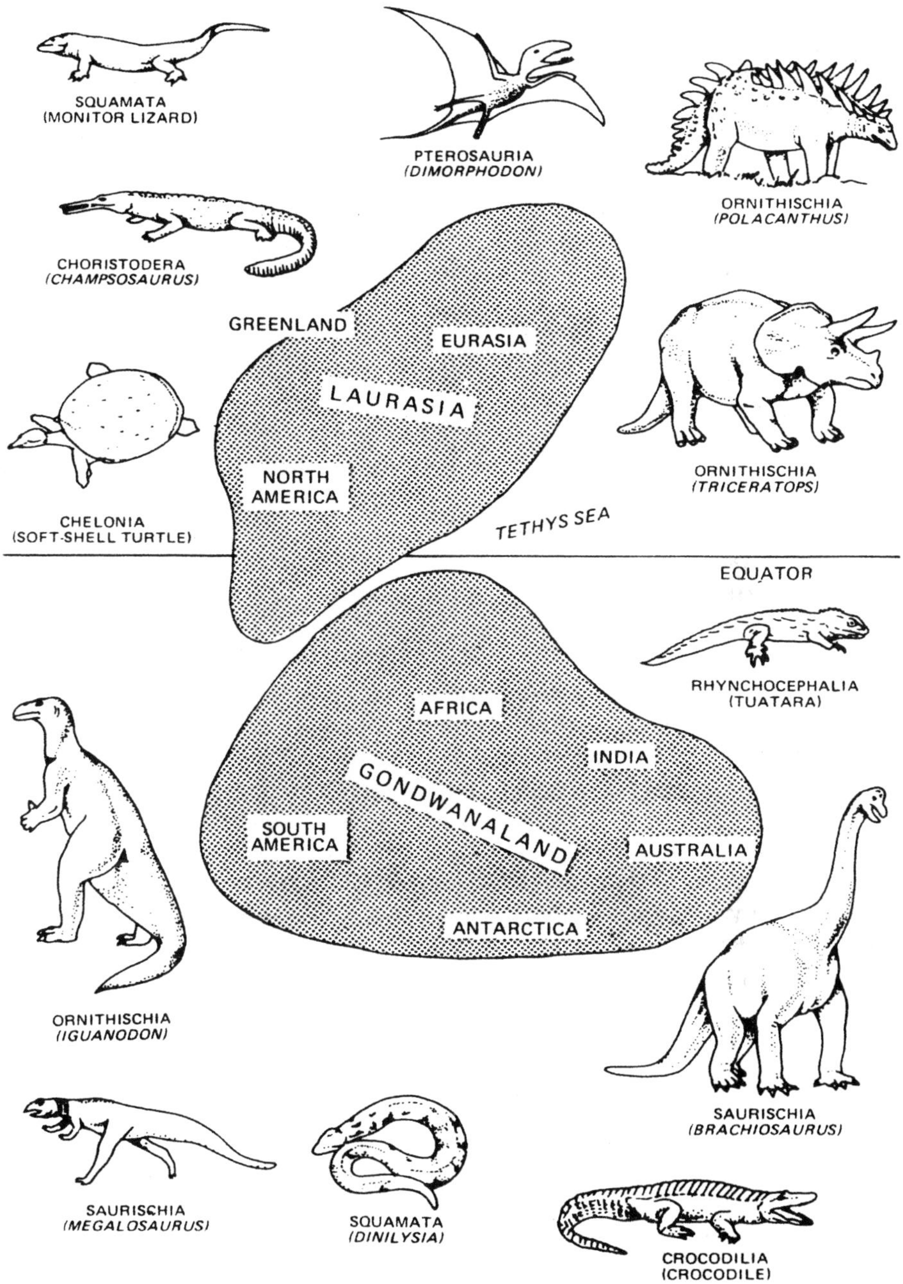

Fig. 84. Laurasia and Gondwanaland showing the 12 major types of reptiles, represented by typical species whose fossils are found in Cretaceous formations. (Cloudsley-Thompson 1977 after Kurtén 1969)

These early dinosaurs were all members of the order Saurischia, which included the suborders Theropoda and Sauropoda (Fig. 2; Table 2). The Ornithischia and Sauropoda had evidently separated before the time of these earliest known dinosaurs. The first massive radiation of the Dinosauria, however, took place in the Upper Triassic period, 225 mya, when or after the Therapsida and Rhynchocephalia had declined (Sect. 8.2).

Throughout the later part of the Mesozoic Era the land mass of the world was divided into the two supercontinents, Laurasia in the north and Gondwanaland in the south (Fig. 84). Most reptilian orders were represented in both of these. Migrations probably took place between them by way of a land bridge in the west where the Tethys Sea was narrowest. Nevertheless, the partial separation explains some of the major differences between the dinosaurs that inhabited the northern and southern continents. During the Upper Cretaceous, these continents gradually separated when Laurasia and Gondwanaland began to break up and, by the end of the Mesozoic, the Atlantic Ocean had become well established – although India was still far from the rest of Asia (see Halstead and Halstead 1981; Benton 1996).

9.2 Dinosaur Weapons

Weapons are of three kinds – those that are used by predators to attack their prey, weapons of defence including armour (Sect. 10.6.4), and agonistic weapons used in intraspecific sexual rivalry. These functions, however, can often be interchangeable; weapons of attack are also employed by carnivores in defensive situations when they are attacked by enemies larger or more formidable than themselves. They may also be used in agonistic combat (Sect. 9.3.1). Again, weapons whose primary function is agonistic may well be used in defence against predators, and so on. The scales of dinosaurs, like those of crocodilians, were embedded and sunk into the thick, protective skin.

9.2.1 Weapons of Attack

The offensive weapons of predatory dinosaurs, including the so-called carnosaurs (Sects. 11.1, 11.3.2), were primarily their sharp, inwardly curved, and serrated teeth (Fig. 85). Especially in the case of bipedal forms, however, the action of these teeth was often supplemented by forelimbs which grabbed the prey while it was being bitten. In numerous cases, curved jaws helped to hold the struggling prey securely. Many predators also had formidable claws that were used to slash the bodies of their prey. For instance, the Early Cretaceous theropod *Deinonychus* ('terrible claw'; Figs. 86, 87) from North America, like other members of the family Dromaeosauridae, was equipped with vicious claws on its hind legs with which it could easily disembowel other dinosaurs. So, too, was *Velociraptor* (Sect. 9.2.2).

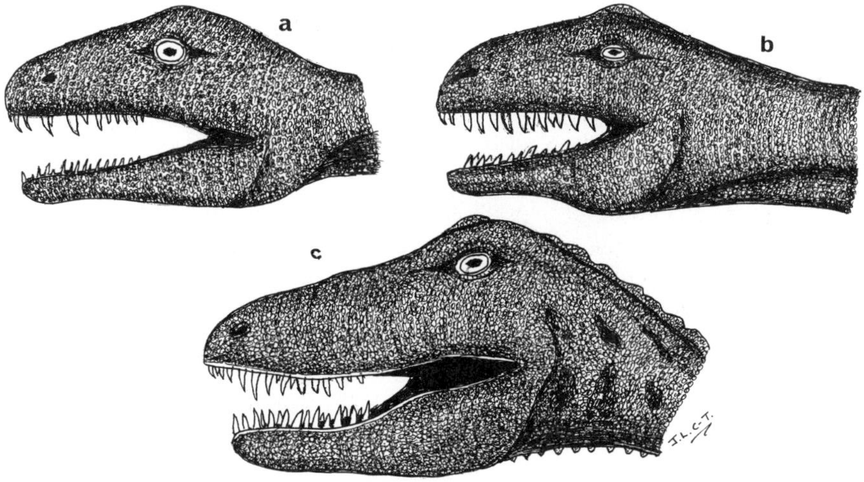

Fig. 85a–c. Carnivorous dinosaur heads (not to scale). **a** *Teratosaurus* (Upper Triassic), **b** *Allosaurus* (Upper Jurassic), **c** *Tyrannosaurus* (Upper Cretaceous).The apparent similarity is the result of convergence in diet and mode of life

The use of weapons by extant animals is invariably supplemented by numerous behavioural and ecological attributes. Similar attributes can only be inferred among fossil forms. For instance, cryptic coloration not only enables potential prey to avoid the attention of predatory enemies, but may also make it possible for the latter to creep upon their victims, unnoticed until the final attack is delivered. By living together in herds, potential prey animals may be relatively safe from attack by predators and also protect their young. Again, by

Fig. 86. *Deinonychus* (Dromaeosauridae; Lower Cretaceous; length ca. 3.5 m). (Cloudsley-Thompson 1999 after Halstead and Halstead 1981)

Fig. 87. *Deinonychus* attacking a herbivorous sauropod. (Adapted from a drawing by Steve Kirk in Brochu et al. 2000)

hunting in packs, smaller predators are able to overcome prey that they could not possibly subdue on their own. Consideration of these and other such matters will be deferred until subsequent chapters, but they must always be taken into account.

9.2.2 Defensive Weapons

The first line of defence among animals that have been detected by a formidable enemy is usually to react by flight. This tactic was undoubtedly employed by most, if not all, of the smaller and more speedy dinosaurs.

Animals that are less agile and cannot move quickly, however, often rely on their armour for defence. This was the case with Mesozoic Testudines, as it is with extant turtles and tortoises. It was also true of the placodonts, the ankylosaurids, nodosaurids and other armoured dinosaurs, as well, to a lesser extent, of the crocodilians which then, as now, had thickened dorsal plates.

Dinosaurs had thick, tough, scaly skin. Their scales were rounded or took the form of tubercles. Large, flat tubercles covered the under surface of the body while, in some ceratopsians and hadrosaurs, the back was coated with rows or clusters of pointed tubercles. Bony plates lay beneath the armoured skin of the ankylosaurs (Ornithischia; Fig. 88) and of the titanosaurid sauropodomorphs in the Cretaceous period.

Anatomical structures that may well have been weapons have been found in the fossils of many species of dinosaurs, but the ways in which they were used remain a mystery. Alexander (1997a) pointed out that no modern animals have weapons like the half-metre spikes on the tail of *Stegosaurus* (Fig. 71), or the heavy bony clubs on the tails of some ankylosaurs (see below). Were these used in defence against predators, in fights between rival males, or both? Probably both!

Fig. 88. *Scelidosaurus* (Scelidosauridae; Lower Jurassic; length ca. 4 m) being inspected by an early carnosaur

Four-legged ornithischians evolved increasingly effective body armour during the Mesozoic Era. *Scelidosaurus* (Fig. 89), one of the earliest and most primitive members of the group, had a small head and toothless beak. However, its massive body was well armoured. The back was covered in parallel rows of bony, spike-studded plates that resembled those of the ankylosaurs, to which it may have been related, and that ran from the neck to the tip of the tail. The stegosaurs (Fig. 71) not only had bony plates or spines on the back, but spines also on the shoulder and tail (see below). We have already seen that the dorsal spines probably had an important thermoregulatory function (Sect. 7.5.2). It is not yet known for certain whether these were covered with horn, which would indicate that they were also defensive weapons, or more probably were covered with a thin layer of vascular skin. The bodies of the ankylosaurids (Fig. 89) were undoubtedly sheathed in bony plates and spines covered with horn, while the nodosaurid

Fig. 89. *Dyoplosaurus* (Ankylosauridae; Upper Cretaceous; length ca. 7.5 m). (Cloudsley-Thompson 1994)

ankylosaurs had long spines on their flanks as well. Finally, many ceratopsians (Fig. 72) bore horns on their nose and brows, while their necks were covered with horny, protective frills.

Claws were used in defence as well as offensively. *Diplodocus* (Fig. 66), for instance, had enlarged thumb claws with which no doubt it could defend itself from the attacks of predatory carnosaurs. These claws were used in addition to its whip-like tail (see below). The claws of predatory dinosaurs would not only have been used for the capture of prey but also in defence against larger enemies.

The tails of dinosaurs were usually held above the ground when the animals moved about. Long, stiffened tails balanced the necks and heads of the bipedal dinosaurs as they ran. Elongated tails also balanced the necks and heads of the four-legged prosauropods and sauropods. Many modern reptiles, especially monitor lizards (Varanidae) and crocodiles, use their tails as weapons. Dinosaurs undoubtedly did the same. The fossilised tail bones of apatosaurs, in particular, often show signs of fracture, doubtless engendered by their use as gigantic whips in defence against predatory theropods. Short, thick tails characterised the four-legged, horned dinosaurs. Like the long tails of the sauropods, these would not only have served as props when their owners reared up to browse, but were also useful as weapons: *Stegosaurus* (Fig. 71) probably jabbed its tail spikes at predatory carnosaurs. Its tail could swing freely because it lacked the bony tendons that stiffened the tails of most ornithischians. *Dyoplosaurus* (Fig. 89) would also have swung its tail sideways at the enemy. Bony outgrowths at the tip of the tail converted this into a formidable club or mace, an adaptation first seen in the sauropod *Shunosaurus*, but better developed among the armoured ankylosaurs (Figs. 90, 116). In these, bones embedded in skin were fused with the tail vertebrae (Lambert 1992).

Some of the most striking developments in defensive weapons evolved among the horned Ceratopsia (Dodson 1996). Of these, the best known species is probably *Triceratops horridus* (Fig. 72). In addition to its heavily armoured collar, it bore three sharp horns on its head. Over a dozen species in the genus *Triceratops* have already been described. Other genera of horned dinosaurs include *Psittacosaurus* (Psittacosauridae), *Protoceratops* and *Breviceratops* (Protoceratopsidae) as well as *Centrosaurus, Brachyceratops, Chasmosaurus* (Fig. 95b), *Anchiceratops, Torosaurus* (Figs. 91, 95d) and *Pachyrhinosaurus* (Fig. 95f) – which was unusual in having a thick pad of bone above the eyes where horns are usually found – to mention but a few of the Ceratopsidae.

Fig. 90a–d. Examples of ankylosaur tail club (not to scale). **a** *Talarurus*, **b** *Euoplocephalus*, **c** *Ankylosaurus*, **d** *Sarchania* (Upper Cretaceous)

Fig. 91. *Torosaurus* (Ceratopsia; Upper Cretaceous; length ca. 8.5 m). (Cloudsley-Thompson 1994)

Fig. 92. Reconstruction of a struggle to the death between a small *Protoceratops* (Ceratopsia; length up to ca. 2.7 m) and a *Velociraptor* (Oviraptoridae; length ca. 1.8 m; Upper Cretaceous). Seen from above; the *Velociraptor* is lying on its side, as found in the fossil record

Fig. 93. *Anchiceratops* (length ca. 6 m) confronting *Albertosaurus* (length ca. 8 m; Upper Cretaceous). (After Dodson 1996)

Fig. 94. *Triceratops* (length ca. 9 m) defeats *Tyrannosaurus* (length ca. 15 m; Upper Cretaceous). (After Dodson 1996)

Ceratopsian weapons may have served not only for interspecific defence but also in intraspecific display and combat (Sect. 9.3.1) – as is seen today among deer and antelope. Nevertheless, not only is there considerable indirect evidence of their use against predatory enemies, but also direct fossil evidence. A life and death struggle between a predatory *Velociraptor* and a small plant-eating *Protoceratops* during the Upper Cretaceous has been preserved for 80 my in the sandstone of the Gobi Desert, Mongolia. The two dinosaurs died while the *Protoceratops* was biting the right arm of the *Velociraptor* which had jammed its left hind killing claw beneath the frill of its prey (Fig. 92). This huge claw had probably penetrated to the base of the neck of the *Protoceratops* where the most important blood vessels were situated. The fight might have been ended by a sandstorm or when a large sand dune, saturated with rainwater, slid onto the struggling animals and smothered them before the *Protoceratops* had been completely overwhelmed. Alternatively, it is possible that the *Protoceratops* was defending its eggs from the *Velociraptor*.

Although ceratopsians have been portrayed as taking up a defensive circle like musk oxen (*Ovibos moschatus*) to protect themselves and their offspring against a large predator, there is as yet no scientific evidence in favour of this hypothesis (Dodson 1996). A dramatic imaginary drawing by Robert Walters (in Dodson 1996) illustrates an *Anchiceratops* confronting the carnosaur *Albertosaurus* in Alberta, Canada, some 70 mya (Fig. 93), while a painting by Wayne D. Barlowe in the same book illustrates the possible consequence of an encounter between *Triceratops horridus* and *Tyrannosaurus rex* (Fig. 94), in which the giant carnosaur is killed by the herbivore (Dodson 1996). Predators did not always come off best in Mesozoic times – any more than they do today! Of course, if *T. rex* was only a scavenger, which might have been the case in view of its diminutive front legs, then it would not have attacked a *Triceratops* in the first place (Chap. 11).

9.2.3
Defensive Display

In many extant reptiles, defensive behaviour is frequently associated with threat and display. However well armed an animal may be, it is always more economical in energy to drive potential enemies away by threatening them rather than by actually indulging in combat. The same must undoubtedly have been the case with the dinosaurs and other Mesozoic reptiles. It is not difficult to imagine the loud hisses and deep roars that would have been emitted by the enormous reptiles of the Mesozoic Era. In addition to warning or aposematic sounds, the display of defensive and offensive weapons would have been enhanced if adorned with brilliant warning colours. Blacks, yellows and reds, in particular, are always extremely conspicuous and would doubtless often have been naturally selected for aposematic purposes in the past, as they are today (Sect. 1.4).

9.3 Reproduction

Among higher animals, reproduction consists of a number of separate phases of activity. Courtship by the male is often preceded by the establishment of a territory and conflict with competing males. This agonistic or aggressive behaviour may consist primarily or even entirely of display (Sect. 9.2.3), and actual combat may be highly ritualised. Agonistic behaviour among dinosaurs probably resulted in the establishment of hierarchies, as it does in extant vertebrates. Interactions between conspecifics may thus result in the avoidance of potentially injurious battles, although fights to the death do sometimes take place (Coombs 1990).

9.3.1 Agonistic Behaviour

Peter Dodson (1996) emphasised that the ceratopsians would have had excellent eyesight and possessed colour vision, as do many modern reptiles including crocodilians. This can be inferred from the fact that visual signals can be found all round fossilised ceratopsian skulls. For example, not only were there conspicuous frills, but these were sculptured with various knobs and processes that would have enhanced an already striking appearance (Fig. 95). These were probably also brightly coloured. Casts of the braincases of many other dinosaur taxa suggest that they, too, would have enjoyed highly developed visual acuity.

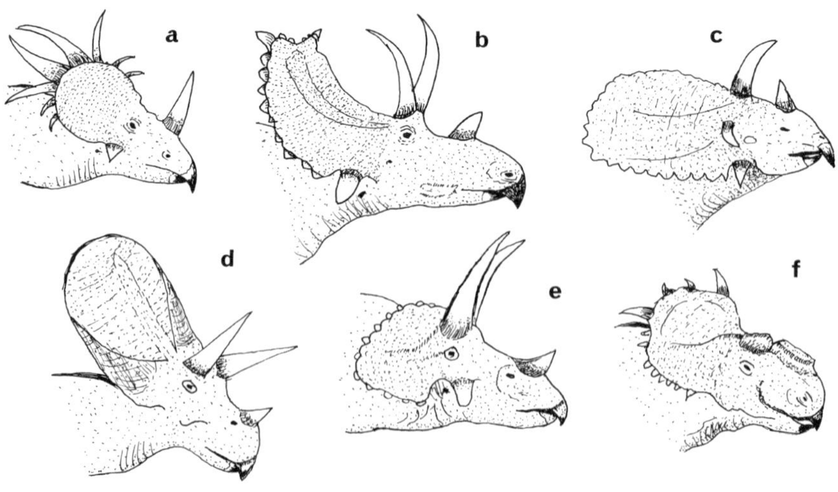

Fig. 95a–f. Typical Upper Cretaceous ceratopsian heads with horns and frills (not to scale). **a** *Styracosaurus* (total body length 5.2 m), **b** *Chasmosaurus* (length ca. 5.2 m), **c** *Pentaceratops* (total body length ca. 6 m), **d** *Torosaurus* (length ca.7.6 m), **e** *Triceratops* (total body length ca. 9 m), **f** *Pachyrhinosaurus* (length ca. 5.5 m)

There seems to be little doubt that *Tyrannosaurus* and the smaller *Nanotyrannus* (length ca. 5 m) from Montana had some degree of overlap in their fields of vision. (This does not appear to have been the case among other carnosaurs so far discovered.) Although overlapping visual fields have been thought to indicate stereoscopic vision, this is not necessarily the case: some birds have strongly overlapping fields of vision yet it has been shown experimentally that they do not possess stereoscopic vision (see discussion in Molnar and Farlow 1990). The same argument could well apply to some other carnivorous dinosaurs such as the ornithomimosaurians and the bird-like Troodontidae.

Some dinosaur horns were indeed dangerous weapons. Their bony cores were often lengthened by keratin sheaths – hard, but weighing less than bone. These might have functioned not only as weapons and in aposematic threat (Sect. 9.2.3) but also in intraspecific display (Fig. 95). Intraspecific display was probably the primary function of the floppy nose bone of *Einiosaurus* from Montana, although this might also have been used in aposematic bluff against potential enemies. So, too, would have been the shield and frill of *Chasmosaurus* (Fig. 95b) from Canada and New Mexico. Unlike the solid bone neck shield and frill of *Triceratops* (Figs. 72, 94, 95e), the shield of *Chasmosaurus* was widely fenestrated. It cannot have been strong enough to have served as protection from predatory attack but, since it was undoubtedly covered with thick skin, it would have been an excellent device for agonistic display, especially if brightly coloured. No doubt its design was the result of a compromise between size, weight and strength. Although usually less dramatic than the horns and frills of ceratopsians (Fig. 95), different types of ornamentation have, not surprisingly, been found in many other taxa of dinosaurs. Its functions usually lie in signalling its possessor's species and in agonistic display (Sect. 10.6.2).

Agonistic behaviour among dinosaurs has been discussed in detail by Coombs (1990) and Sampson (1997) among others. According to the well-documented account by Coombs, the dinosaurian features reasonably to be interpreted as visual or vocal threat display – and which could sometimes also have been actual weapons – included:

1. Bright colours: the presence of these can be inferred from the excellent colour vision of crocodiles and birds, as well as from the presumably well-developed bird-like eyes of dinosaurs.
2. Flamboyant tails: dinosaur tails could have been used for display, especially when unusually long as in Diplodocidae (Fig. 66) or decorated with armour plates, as in stegosaurs (Fig. 71) and ankylosaurs (Fig. 90). Tails are used for display in many modern vertebrate taxa including reptiles, birds and mammals. The tails of most dinosaurs were carried clear of the ground, as already mentioned, and tail 'drags' are seldom to be seen in association with fossil trackways. Some genera clearly used their tails as weapons (Sect. 9.2.2).

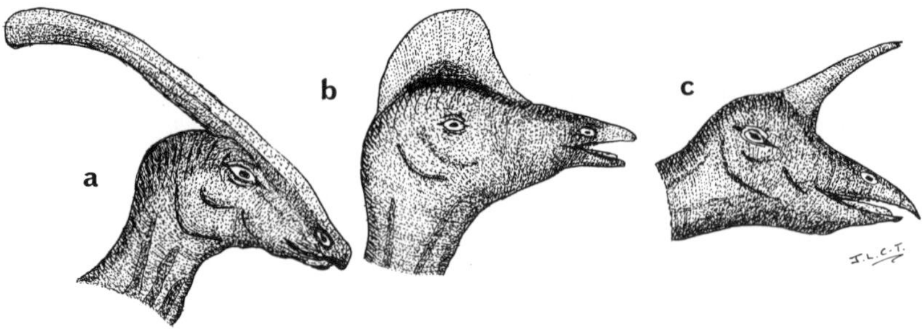

Fig. 96a–c. Upper Cretaceous crested hadrosaur heads (not to scale). **a** *Parasaurolophus* (total body length ca. 10 m), **b** *Corythrosaurus* (total body length ca. 9 m), **c** *Tsinatasaurus* (total body length ca. 10 m)

Fig. 97. *Iguanodon* (Iguanodontidae; Lower Cretaceous; length ca. 9 m) shown here in an upright posture. (After Augusta 1956)

3. Upright posture, long necks: sauropods might well have indulged in giraffe-like sparring with their long necks: they probably reared up on their hind legs, propped by the tail, when they were using their large claws in combat. Some genera (e.g. *Brachiosaurus*; Fig. 68; and *Camarasaurus*) might have had an inflatable proboscis for display purposes – like those of elephant seals (*Mirounga* spp.) – but this hypothesis has been disputed.
4. The dorsal plates, flattened bodies and elongated hind legs of the stegosaurs would have been very suitable for lateral display, while the tail spikes were undoubtedly used in combat (Sect. 9.2.2).
5. The crests and complex narial passages of the duckbilled hadrosaurs (Fig. 96) were almost certainly used for visual and sometimes vocal display. The thumb spikes of *Iguanodon* (Fig. 97) might possibly have been used both for display and combat, as might the canine tusks of the vegetarian *Heterodontosaurus* (Fig. 98).
6. Ankylosaurids could have used their tail clubs for combat, and their broad, triangular skulls for thrusting head to head in competition for dominance (Fig. 99).

Fig. 98. *Heterodontosaurus* (Heterodontidae; Lower Jurassic; length ca. 90 cm). (After Norman 1985)

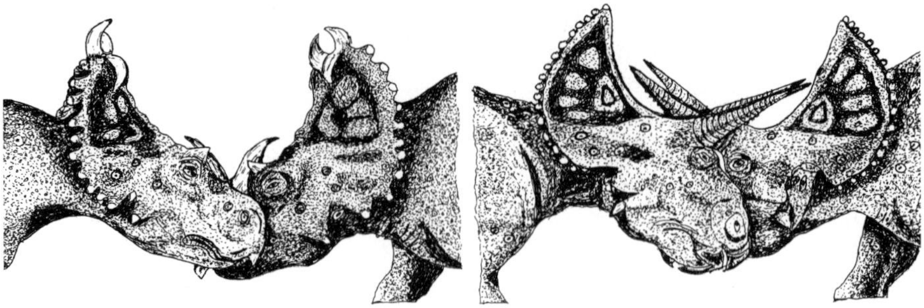

Fig. 99. Possible styles of grappling combat in *Centrosaurus* (*left*) and *Triceratops* (*right*). (Based on a drawing by Bill Parsons in Sampson 1997)

7. The frills and horns of the ceratopsians (Fig. 95) were probably used for both display and combat, as we have seen. Fossil skulls with punctured wounds on the frills and damaged horns are proof that the animals indulged in tests of dominance.
8. It is generally accepted that the thickened crania of *Stegoceras* and other Pachycephalosauria were an adaptation for sexual combat (Fig. 100; Galton 1970b).
9. The sabre-like claws of *Deinonychus* (Figs. 86, 87) and other Dromaeosauridae might well have been used in sexual combat like the spurs of fighting cocks, their reinforced tails forming a tripod with the rear legs when the animals were fighting.
10. Various unrelated genera such as the iguanodont *Ouranosaurus* (Fig. 101a) from Niger, the duckbilled *Hypacrosaurus* (Fig. 101b) from North America, as well as the carnosaurs *Acrocanthosaurus* (Fig. 101c) from North America and *Spinosaurus* (Fig. 101d) from North Africa possessed sail-like extensions supported by elongated neural spines, which were almost certainly used for display.

Fig. 100. *Stegoceras* (Pachycephalosauria; Upper Cretaceous; length ca. 2 m) in sexual combat. (Cloudsley-Thompson 1994 after Halstead and Halstead 1981)

Fig. 101a–d. Dinosaur display structures supported by neural spines. **a** *Ouranosaurus* (Iguanodontidae; Lower Cretaceous; length ca. 7 m), **b** *Hypacrosaurus* (Spinosauridae; Upper Cretaceous; length ca. 9 m), **c** *Acrocanthosaurus* (Spinosauridae; Lower Cretaceous; length ca. 12 m), **d** *Spinosaurus* (Spinosauridae; Upper Cretaceous; length ca. 12 m). (After Palmer 1999)

Fig. 102. Theropods with cranial crests. *Left Dilophosaurus* (Megalosauridae; Lower Jurassic; length ca. 6 m). *Right Ceratosaurus* (Ceratosauridae; Upper Cretaceous; length ca. 6 m)

11. Finally, many carnivorous dinosaurs would have displayed their formidable teeth with open jaws. Several genera of carnivorous dinosaurs had cranial crests that would almost certainly have been brightly coloured and used for display. Examples include *Dilophosaurus*, *Ceratosaurus* (Fig. 102), and possibly *Ornitholestes*, all from North America.

9.3.2 Courtship and Mating

In animals living today, courtship and mating involve a whole range of visual, acoustic, tactile and chemical signals. Fossils, by their very nature, can only provide unequivocal evidence for the first of these. It is generally assumed, however, that many of the morphological structures and ornaments, such as horns, frills, crests, spikes, thickened skulls, bosses, large teeth, big eyes and so on (Sect. 9.3.1), were not only used in combat but also in eliciting behavioural responses (Horner 1997).

Adaptations are seldom the response to a single factor, as already emphasised. In most cases, several selective factors are involved. For example, frills and crests may have been exploited in both combat and visual display – when their presumed bright colours would have been important (Sect. 1.4). These adornments could also have played a role in cooling the brain. If the internal surfaces were lined with moist membrane, evaporation from this would have reduced the temperature of the surrounding tissues. Some of the Late Cretaceous duckbilled hadrosaurs had small, solid crests, but in others the crests were larger and more conspicuous. Specimens with larger crests were probably males, those of apparently the same species with smaller crests quite likely females. In such cases (Sect. 2.3.1), almost certainly both visual and vocal display would have been used. In *Parasaurolophus* from North America (Fig. 96a), for example, the hollow crest was probably used as a vocal resonating chamber. It would have produced a sound quite different from that of, say, *Corythrosaurus* (Fig. 96b) also from North America, and other genera and species of duckbilled hadrosaurs. Females that had shorter crests must have emitted sounds pitched higher than those of males of the same species. The suggestion has also been made that the hollow crest of *Corythrosaurus* might have enhanced the animal's sense of smell, thereby helping it to find food, detect approaching enemies and keep up with the herd. *Tsintasaurus* (Fig. 96c) from China had a horn on its head between the eyes. This gave it a bizarre unicorn-like appearance. If a flap of skin were anchored between the tip of the horn and the beak, it could have been inflated for use as a signalling device in courtship, as a threat to rivals, or both. Alternatively, if the horn was mounted incorrectly in the original specimen and actually pointed backwards like that of *Parasaurolophus* (Fig. 96a), it would have served the same purpose as that of other hadrosaurs and acted as a vocal resonator. This really seems to be the most likely explanation!

Ornamentation in fossil vertebrates has been defined by Vickaryous and Ryan (1997) as "any unique or modified anatomical feature (whether skeleton or preserved/inferred soft-tissue) that suggests a use in social behaviour, including taxon recognition (inter- or intraspecific), gender identification, and/or the recognition of the state of sexual maturity." Many types of ornamentation have been described in the previous section; others are listed by Vickaryous and Ryan (1997). Details of mating behaviour can only be inferred by comparison with the known behaviour of extant reptiles and birds.

9.3.3
Nesting Behaviour

Probably most, if not all dinosaurs, were oviparous – but this supposition has been confirmed only in a comparatively small number of taxa. Those examples that are known and have been described include *Troodon* (Theropoda), *Oviraptor* (Theropoda), *Maiasaura* (Ornithopoda), and *Hypacrosaurus* (Ornithopoda) among others. All of these date from the Upper Cretaceous and all were discovered in North America – except for *Oviraptor*, which was found in southern Mongolia (Horner 1997). Prosauropod, sauropod, ornithopod and ceratopsian eggs of less well known provenance, however, have been found in almost every continent (Lambert 1992).

Dinosaur eggs had hard shells, but they could not have exceeded about 7 kg in weight. They were therefore very small in comparison with the sizes of the dinosaurs that produced them. For instance, the eggs of the sauropod *Hypselosaurus* are among the largest dinosaur eggs known, yet they measured only 30.5×25.5 cm (Fig. 103). They were, therefore, comparable in size with the eggs of *Dinornis maximus* – the largest of the moas – and of the elephant bird *Aepyornis titan* of Madagascar, which measured 33×24 cm. These eggs have been described as the largest single cells to have evolved in the animal kingdom!

Although the shell of an egg may be hardened to protect the developing embryo within, it must remain sufficiently porous to allow for the passage of air from the outer world. It cannot, therefore, be inordinately thick or it will prevent respiration of the embryo from taking place. Moreover, it cannot be excessively hard or it will prevent the young animal from hatching. Since the thick-

Fig. 103. Comparative sizes of the eggs and adult of *Hypselosaurus* (Titanosauridae; Upper Cretaceous; length ca. 12 m) The eggs would normally have been buried in the nest (see text)

ness of eggshells is limited in this way, there is a geometric limit to the sizes of eggs. This limit is approached when eggs reach such volume that the pressure of the internal fluid exceeds the strength of the shell that contains it (Colbert 1962b).

It seems probable, therefore, that the eggs of *Aepyornis titan* and of *Hypselosaurus priscus* were at about the maximum size to which any terrestrial eggs, even those of *Apatosaurus* or *Brachiosaurus,* could reach. Consequently, even the largest of the dinosaurs would have started life as comparatively tiny hatchlings. The same occurs in the largest of modern reptiles, such as the giant saltwater crocodile (*Crocodylus porosus*) and the great leathery turtle (*Dermochelys coriacea*). The rate of growth in dinosaurs from egg to adult must have been impressive. A newly hatched *Hypselosaurus* weighed about 0.9 kg and grew to an adult that weighed perhaps 10 tonnes. Consequently, there must have been an increase in weight in the order of not less than 10,000 times. In *Brachiosaurus* spp. there may have been an increase during life of about 50,000 times! In view of the relatively small sizes of their offspring, it is to be expected that many of the dinosaurs would have tended their eggs and young, as most birds do. Colonial nesting was probably not uncommon, and dinosaurs evidently had favoured nesting sites. Some of the larger dinosaurs, especially sauropods, may well have been ovoviviparous.

Different species of dinosaur laid rounded or oval eggs on bare ground, in sandy hollows, or in nests with raised mud walls. Known dinosaur nests and eggs exhibit surprising diversity in their forms. Those of sauropods found in Europe and the Gobi Desert of Mongolia were large and subspherical in shape with a highly developed system of pores in the shell. The orifices of the pores occupied about half of the egg surfaces. These eggs probably developed in moist substrates, not far from water. The eggs of hadrosaurs, hypsilophodonts, protoceratopsians and theropods, on the other hand, were laid in nests composed of soil and vegetation, like those of alligators and some crocodiles. The eggs of hadrosaurs were rounded, whereas those of protoceratopsians, hypsilophodonts and theropods were elongated (see relevant entries in Currie and Padian 1997).

Each female *Protoceratops* laid a dozen or more eggs, sometimes as many as 18, in a sandy hollow where they formed a spiral with the narrow ends pointing inward (see below). Apparently several females sometimes laid their eggs communally in the same nest. The adults of colonial species might well have combined to protect the eggs and young from marauding theropods, forming an outwardly facing ring to shield them as musk oxen deter wolves today, as already mentioned (Sect. 9.2.2; Lambert 1992).

Although there is no evidence to suggest that dinosaurs were other than invariably oviparous, and large numbers of fossil eggs have been unearthed, very few can be shown to have been laid by any particular species. A notable exception, however, is the discovery in Mongolia of a fossil theropod *Oviraptor* lying on a nest which contained about 22 eggs that had probably been overwhelmed by a sand storm. On the other hand, the animal in question could have been robbing the nest of another dinosaur at the time of its death. *Oviraptor* lacked

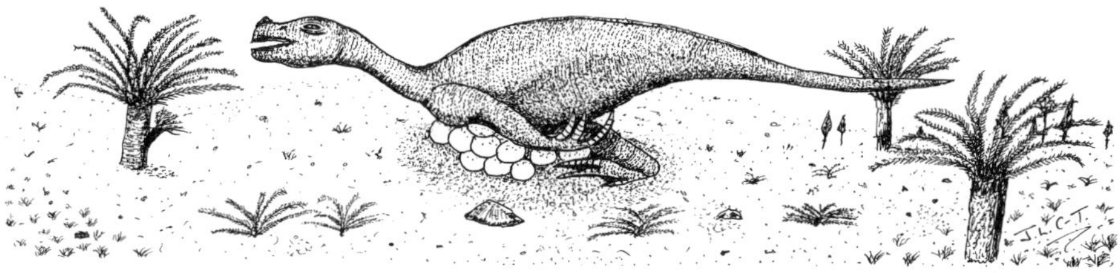

Fig. 104. *Oviraptor* (Oviraptoridae; Upper Cretaceous; length ca. 1.8 m) on its nest shortly before death. (Based on Norrell et al. 1995)

teeth for biting prey, but two short teeth, pointing down from the roof of its mouth, may have been used to open the shells of other dinosaurs' eggs in the same way that a sharp vertebral process from the neck enables egg-eating snakes (*Dasypeltis* spp.) to cut through the shells of the birds' eggs they swallow. It appears more likely, however, that the Mongolian *Oviraptor* was incubating its own eggs because the front limbs were directed posteriorly, with both arms wrapped around the nest (Fig. 104). Another possibility is that the animal perished whilst in the act of ovipositing, not incubating. This seems to be obviated, however, both by the absence of eggs within its body cavity and by the neat arrangement of the eggs (Norrell et al. 1995). The later discovery of an egg containing a developing embryo proved that the *Oviraptor* had not been robbing the nest of another species of dinosaur (Chiappe and Dingus 2001). The fact that the eggs were comparatively large, were deposited in pairs and partly buried, indicates a degree of environmental heating. The position of some adult dinosaurs with the small forelimbs draped over their eggs, suggests that they possessed a degree of insulating pelage (Paul 1997).

Among the oldest and smallest dinosaur eggs yet discovered are some that measured only 25 mm on the longest axis, while the skeletons of the babies found with them did not exceed 20 cm. These animals from the Upper Triassic of Argentina were named *Mussaurus* (mouse lizard). Since no other specimens have been found, the size of the adults is not known, but it is generally believed that they could have grown to a length of 3 m with body proportions typical of prosauropods. The babies had large heads and short necks as are typical of very young animals. They clearly could not have survived without parental care. There is evidence, therefore, that the young both of hadrosaurs and of prosauropods were fed and looked after by their parents. So, too, were the young of *Protoceratops* (Fig. 105), already referred to, whose eggs were about 25 cm long with thin, wrinkled shells. Some of the eggs, found in the Upper Cretaceous of Mongolia, contained fossilised fragments of embryonic bones.

The care of eggs and young of dinosaurs probably ranged from nil at one extreme to the feeding and guarding of the young at the other. Eggs laid in simple holes were probably abandoned after being buried, whereas open nests were more likely to have been guarded. The incubation temperatures of nests in mounds of fermenting vegetation may well have been regulated by the parents,

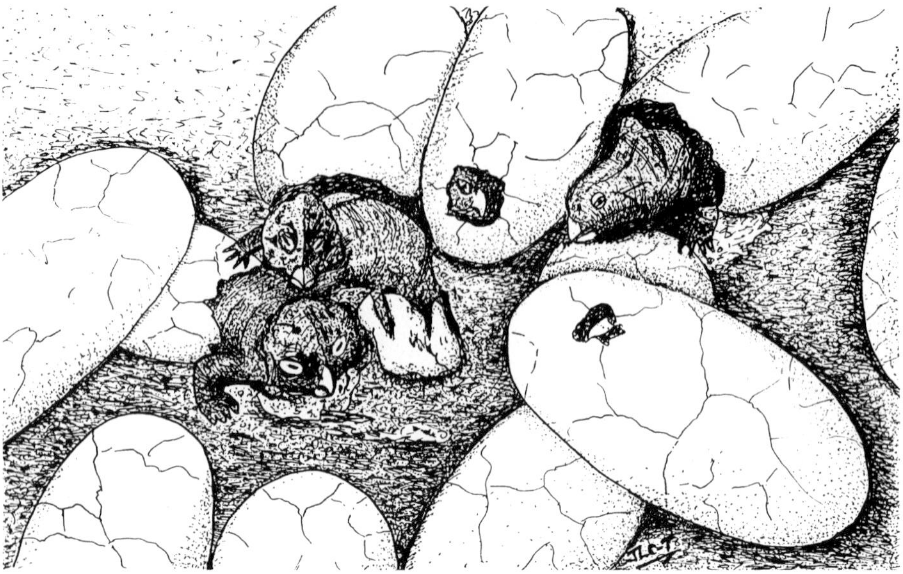

Fig. 105. Clutch of *Protoceratops* hatching eggs (Protoceratopsidae; Upper Cretaceous). (Cloudsley-Thompson 1994)

by adding or removing nest material as required. This behaviour is found in megapode birds today. In contrast, the extreme size of the largest dinosaurs would probably have rendered any form of brooding impossible. Very large numbers of eggs would then have been laid in each clutch, however (Sect. 7.3.4). Post-nestling dinosaurs may well have lived together in groups, possibly on their own or perhaps guarded by their parents. When they grew larger they would have joined adult herds without the risk of being trampled to death. Hadrosaurs built mound nests in colonies that were probably inhabited by several successive generations. This is indicated by the large number of broken egg shells found in such places.

It has been argued that the ossified pelves of baby hadrosaurs indicate that they enjoyed well-developed locomotor ability and consequently had precocial habits. On the other hand, the shafts of the leg bone were so poorly ossified that the babies might have been immobile and therefore altricial (helpless at birth). The subject has been reviewed by Paul (1997) who suggested that hadrosaur nestlings might have been semi-altricial with just sufficient locomotor ability to flee from their nests if a predator penetrated the defences of their parents. When provided with large quantities of food by their relatively enormous parents, the tiny juveniles would have grown rapidly and soon become large enough to move in the company of adults without the risk of being inadvertently trodden on. Hadrosaurs probably foraged over considerable distances and would thus have been able to obtain vegetation when grazing had reduced to desert the region around the nesting ground.

The reproductive strategies of dinosaurs, like those of large birds and mammals today, embraced rapid growth and high rates of population expansion. Mammals differ, however, in that their rates of reproduction are low; but mammalian young benefit from lactation (Paul 1997).

9.3.4 Growth and Development

According to Paul (1997), dinosaurs would have grown extremely rapidly. The larger the adults, the greater the number of eggs that they would have laid – probably many tens of thousands during the life spans of the largest sauropods. The extreme ratio between size on hatching and at adulthood would have necessitated very rapid growth for sexual maturity to have been reached within two or three decades. Breeding would have had to begin within that time scale if sufficient juveniles were to survive to sexual maturation. Dinosaurs could not have lived and grown for more than 100–150 years at most. Smaller dinosaurs may have been K-strategists, with low birth rates and advanced parental care, although with fast growth rates. Large species, on the other hand, were undoubtedly fast-breeding *r*-strategists, with high levels of egg deposition, fast growth and high juvenile mortality. Only small numbers of juveniles needed to reach sexual maturity to build up and maintain large populations. These would have been skewed more towards juveniles than were those of the K-strategists with lower birth rates and the advantage of parental care.

10 Herbivorous Dinosaurs

10.1 Introduction

Many taxa of dinosaurs evolved into successful herbivores. Of these, the principal groups were the basal Sauropodomorpha or prosauropods of the Late Triassic, the Sauropoda, and the Ornithischia. The most important taxa of ornithischian dinosaurs were the Fabrosauridae, Heterodontosauridae, Hypsilophodontidae, Iguanodontidae, Hadrosauridae, Pachycephalosauridae, Stegosauria, Ankylosauria and Ceratopsia. Some of these were bipedal and will be dealt with after the quadrupedal forms.

If it did not feed upon the eggs of other dinosaurs (Sect. 9.9.3), the theropod *Oviraptor* could well have fed upon plant matter because its jaw muscles – as indicated by the bones of the skull – must have been rather similar to those of ceratopsians. Perhaps it ate both. In contrast to those of modern reptiles, the skulls of herbivorous dinosaurs were adapted in a number of ways for chewing vegetation. A transverse power stroke was achieved by different methods in the various taxa for grinding food. Another feature of the versatility of dinosaurs was the ways in which they exploited all the positions in the mouth processor/gut processor continuum (King 1996), as explained below.

10.2 Teeth, Gastroliths and Digestion

Prosauropods such as *Plateosaurus* (Figs. 5, 65, 83e) and *Massospondylus* (Figs. 65, 83f; Sect. 7.3) had lightly built skulls and the jaws were articulated below the line of the teeth. Consequently, the teeth of the upper and lower jaws could be almost parallel as they approached each other when the mouth was closed. In addition, the biting force would have been distributed more evenly than it is when rows of teeth approach each other in a scissor-like fashion (King 1996). Galton (1985) demonstrated convincingly that they were herbivores. As noted by John Attridge and his colleagues in 1985, *Massospondylus* appears to have possessed a gastric mill. They suggested that comparatively large stomach stones or gastroliths were probably lodged in the muscles of the stomach walls in a region which functioned like an avian gizzard to grind the food. Unlike many ornithischian dinosaurs, the prosauropods did not have complicated teeth adapted for chewing: they were gut processors. To maintain their large bulk, these animals must have had to spend much of their time feeding, so their gastroliths would have become rounded and polished in a comparatively short space of time. They would then have been regurgitated and fresh, sharp-edged stones swallowed in their place. The lack of wear on the teeth of most prosauropods suggests that the teeth of the upper and lower jaws did not come into contact with each other. Oral processing probably involved a piercing and puncturing action of the teeth before the food was ingested and the gastric mill took over (Sect. 11.2).

Stomach stones were also swallowed by sauropods but, in the absence of fossil evidence regarding the soft tissues, the function of these, like those of prosauropods, cannot necessarily be equated with that of grit in the gizzards of birds, as is often presumed to be the case. Indeed, Gillette (1994) questioned the notion of any direct participation of dinosaur gastroliths in grinding vegetable food. From his study of the stomach stones associated with *Seismosaurus* (Fig. 108), he argued at some length that they probably served to stir the digestive juices of the animal. Although usually rounded, some were more polished than others, suggesting that they had been swallowed somewhat earlier. None was angular with sharp edges. This indicated that some gastroliths remained in the digestive tract for long periods – even years – until they were passed out in the faeces with the remnants of digested food. A problem remains: how were the polished stones selected by the gut? Maybe when stomach stones, both worn and angular, were excreted by chance, they were replaced only by gastroliths with sharp edges. This would account for some very worn stones being found in the alimentary canal.

The gastroliths of *Seismosaurus* were found in two groups, one large cluster in the forward part of the alimentary canal, and a smaller one in the region of the pelvic bones. These were insufficient in number to have served as a grinding device for pulverising food in a crop gizzard, but might have helped fermentation by mixing digestive juices with the plant material that had been swallowed. The second cluster of gastroliths probably served the same function at the lower end of the digestive tract (Gillette 1994).

Tiffney (1997) summarised the various opinions that have been expressed. Gastroliths could have functioned to aid digestion either by muscular activity in the stomach causing them to rub against each other and thereby crushing and grinding the food; or they might have served to mix the contents of the alimentary canal, thus ensuring more complete digestion. Although there is no evidence for the presence of a gizzard, muscular crop or muscular stomach in dinosaurs, King (1996) argued that there is some indirect evidence for one. She quoted the review of Farlow (1987) who presented five scenarios to account for how sauropods, ankylosaurs and stegosaurs, in particular, could have managed to obtain sufficient energy by gut processing. First, they might have had lower standard metabolic rates than dinosaurs such as hadrosaurs and ceratopsians and, consequently, did not require the batteries of sophisticated teeth found in those families. Alternatively, they had lower metabolic requirements because they led less active lives. Thirdly, they might have fed on low-fibre, highly nutritious food. Fourthly, the enormous size and relatively lower metabolic rates of sauropods may have reduced rates of turnover so that the food could be digested more slowly and, finally, sauropods, ankylosaurs and stegosaurs might have had lower digestive efficiency but higher rates of turnover. Whatever the case, herds of feeding dinosaurs, to use the words of Benton (2004) "must have rattled, grunted and burped furiously as their rough plant diet was reduced to a digestible state!" This subject has also been discussed by Norman (1985, 1991), Dodson (1990, 1996) and Fastovsky and Weishampspel (1996), among others.

10.3 Basal Sauropodomorphs – Prosauropods

Despite their similarity in appearance to the Sauropoda (Sect. 10.4) which radiated much later, in the Upper Jurassic and Cretaceous periods, the sauropodomorphs of the Upper Triassic were not related to them (Sect. 9.1). The apparent similarity must have resulted from parallel evolution which took place in similar environments at different periods of earth history. *Plateosaurus* (Figs. 65, 83e) is probably the best known of the prosauropods. Many well-preserved skeletons have been excavated from Upper Triassic sandstone beds throughout western Europe. The presence and accumulation of fossils can be explained by one of two hypotheses. According to the first, the animals travelled together in large herds from one feeding ground to another, and occasionally a whole herd perished as the result of some misfortune such as a landslide, falling over the edge of a cliff, or becoming trapped in a marsh. Alternatively, individual corpses might have been washed away and piled up together by periodic flash floods in a semi-arid environment.

Plateosaurus was much larger than its plateosaurid relative *Massospondylus* (Fig. 83f). It probably walked on all fours for most of the time, as already noted, feeding on the foliage of conifers, cycads and other trees and shrubs that flourished during the Late Triassic (Sect. 9.1). *Massospondylus* had especially large hands which would have been useful for grasping food plants. In contrast, the tiny *Mussaurus* (mouse lizard) from Patagonia was entirely quadrupedal, although it also belonged to the family Plateosauridae. It was originally believed not to have exceeded 20 cm in length, but the specimens from which the genus was described were very young, and palaeontologists now think that the adults could have reached 3 m.

Another family of basal sauropodomorphs, the Melanorosauridae, contains the largest of the prosauropods. One of these, *Riojasaurus* (Fig. 106) from South America, was up to 11 m long and much more heavily built than were any of the plateosaurs. Other large melanosaurids that lived in southern Africa at about the same time were *Roccosaurus* and *Thotobolosaurus* of the Upper Triassic, and *Vulcanodon* from the Lower Jurassic. These animals had to walk on four legs in order to support their weight, and their skeletons were modified accordingly. *Vulcanodon* from southern Africa showed a mixture of prosauropod and sauro-

Fig. 106. *Riojasaurus* (Melanosauridae; Upper Triassic–Lower Jurassic; length ca. 11 m). (Based on Palmer 1999)

pod characters and could have linked the two infraorders (Palmer 1999). Benton (2004) included it with the Sauropoda. The ancestry of the sauropods is still obscure, although the traditional assumption that they evolved from bipedal prosauropods was questioned as long ago as 1965 by Charig, Attridge and Crompton, to whose paper the reader is referred for a review of earlier opinions on the subject.

In general, members of the family Plateosauridae were mostly relatively agile bipedal and quadrupedal reptiles that could browse on tall trees with the aid of their long necks and hind legs. Their teeth were coarsely serrated and shredded plant fibres which were then pulped or stirred by the gastric mill. They probably lived in herds that afforded some protection from predatory theropod dinosaurs (Chap. 11) and defended themselves with the large claws on their hands. The melanorosaurids, in contrast, may have been relatively invulnerable to predators when full grown on account of their large size. They had small heads on long, flexible necks so that they could feed at ground level and also reach up into trees. The suggestion has been made that, as food plants became scarce on the ground when the world dried up at the end of the Triassic, long necks were advantageously selected (Brett-Surman 2000).

10.4
Sauropods

The sauropods were gigantic quadrupedal, herbivorous dinosaurs. Their large size may have been facilitated by increased oxygen content of the atmosphere, which also contributed to the evolution of aerial reptiles (Sect. 6.3). They appeared in the Lower Jurassic and survived until almost the end of the Cretaceous period (McIntosh et al. 1997). During most of this time, they were the dominant terrestrial herbivores of the earth. First thought to be amphibious, they are now recognised as having been facultative bipeds. They would have used a quadrupedal stance when moving slowly, but adopted a bipedal position for fast locomotion or when browsing on high vegetation. Even in the early stages of their evolution, they were enormous (Sect. 7.4). Most sauropods were well over 15 m in length. Among the largest was *Paralititan* (Fig. 107) from Bahariya oasis in the Sahara Desert of Egypt. With a humerus measuring 1.69 m in length, *Paralititan* must have been ca. 30 m long and could have weighed between 75 and 80 tonnes. Even heavier, *Argentinosaurus* is estimated to have been 7.5% longer still. When first discovered, *Seismosaurus* ('earth-shaking lizard'; Fig. 108) was claimed to have exceeded a length of 33 m (Gillette 1994). In general, however, the longest sauropods were not necessarily the heaviest. The graceful *Diplodocus* (Fig. 66) reached a length of ca. 25 m, but weighed only about 20 tonnes. In contrast, the massive body of *Barosaurus*

Fig. 107. *Paralititan* (Diplodocidae; Upper Jurassic; length ca. 30 m)

Fig. 108. *Seismosaurus* (Diplodocidae; Upper Jurassic; length ca. 33 m). (Based on Gillette 1994)

('heavy lizard'; length ca. 27 m) from East Africa and North America probably weighed 30 tonnes – or even more. This huge dinosaur may have been only slightly longer than *Diplodocus* because, although its neck was extraordinarily elongated (ca. 9 m), its tail was relatively short.

The division Neosauropoda contains five families: Cetiosauridae, Brachiosauridae, Camarasauridae, Titanosauridae (Fig. 103), Nemegtosauridae and Diplodocidae (Benton 2004). The largest dinosaurs of all were members of the Diplodocidae – *Paralititan* (Fig. 107), *Argentinosaurus, Seismosaurus* (Fig. 108), *Apatosaurus* (Fig. 109; see below) and so on – or of the Brachiosauridae. *Brachiosaurus* (Fig. 110) from Europe, Africa and North America is the largest of all land animals for which a complete skeleton exists. Mounted and displayed in the Palaeontological Museum of Humboldt University, Berlin, its bones were excavated in Tanzania in 1908–1912. Brachiosaurs differed from other sauropods in that their front legs were longer than their hind legs. Consequently, the body sloped backwards like that of a giraffe. The relatively short tail resulted in an overall length of only ca. 25 m. Calculations regarding their mass range between 50 and 80 tonnes. The Berlin specimen reached 22.5 m in length and its head was

Fig. 109. *Brachiosaurus* (Brachiosauridae; Upper Jurassic; length ca. 23 m). (Based on Špinar 1995)

held about 13 m above ground level. Gillette (1994) mentioned the bones discovered in western Colorado by J.A. Jensen and described in 1985 as representing other supergiant dinosaurs – *Supersaurus, Ultrasaurus* (based mostly on a single vertebra) and *Dystylosaurus*. Two of these, however, could have belonged to the same individual, while at least one of them might be part of the skeleton of a *Brachiosaurus*.

The long-necked browsing Jurassic sauropods were characterised by a uniform body plan as far as their post-cranial skeleton was concerned. Their skulls were small with the nostrils opening dorsally; they had straight limbs

Fig. 110. A herd of *Apatosaurus* (Diplodocidae; Upper Jurassic; length ca. 21 m). (Cloudsley-Thompson 1994)

and broad feet like those of elephants. Pads on these would have absorbed the shock of each footfall. Their long, supple necks were balanced by long tails. The teeth of *Diplodocus* and *Apatosaurus* were reduced in number and limited to the front of the jaws. Teeth were more numerous in *Brachiosaurus* and *Camarosaurus*, however, where they were present on the sides as well as at the front of the jaws. In the latter genus, they were relatively large and robust, suggesting that some mastication of the food may have taken place – although most of it must have been processed in the gut (Sect. 10.2; Colbert 1993). Lacking the ability to chew fodder, as their peg-like teeth did not occlude, it seems highly likely that the large sauropods closed their mouths around the branches of trees and pulled their heads back, raking the foliage off as they did so (McIntosh et al. 1997; Tiffney 1997).

Fossils of many of the largest dinosaurs have been found in what are now South America and Africa. As explained above (Sect. 9.1; Fig 84), during much of the Mesozoic Era, Gondwanaland, the supercontinent in which they lived, was partially separated from Laurasia in the north by the Tethys Sea. Titanosauridae, for instance, the latest family of sauropods to evolve, survived until about the end of the Cretaceous period. Although a few genera have been found in southern parts of Laurasia, they were a truly Gondwanan group. Sauropods were nearly extinct in the rest of the world when titanosaurids were the dominant plant-eaters in Gondwanaland, occupying the ecological niche of the ornithopods in Laurasia. Some meat-eating dinosaurs, such as *Carnotaurus* and other members of the family Abelisauridae, were also typically Gondwanan. A few of them, however, must have been able to cross the Tethys Sea (Sect. 9.1) and radiate on a small scale in various parts of Laurasia. This subject has been accorded fascinating treatment by Benton (1996).

Evidently some dinosaurs not only aggregated for feeding and breeding purposes, but remained together throughout much of their lives (Ostrom 1972), so they were probably truly social animals (Figs. 109, 110). The herds might well have had home ranges, but it seems unlikely that these would have been defended against conspecific herds. Nevertheless, hierarchical behaviour may have occurred – especially among the duck-billed hadrosaurs that had large crests but were without elongated nasal bulbs (Sect. 9.3.1). Probably most species, especially among the large sauropods, were not aggressive. Odours and colour may well have been exploited in sexual displays.

10.5 Herbivorous Bipeds

Of all the Ornithischia (Table 2), the Ornithopoda showed the greatest tendency towards bipedality. Like all ornithischians, they were plant-eating dinosaurs. In most cases, their cheek teeth were slightly inset from the edges of the jaws. The spaces outside were probably enclosed by fleshy cheeks which prevented food from falling out while they were chewing it. It seems quite possible that the evolution of cheeks may have been responsible for the outstanding success of small and medium-sized herbivorous dinosaurs during the Jurassic and Cretaceous periods. Linked to this was the failure to compete and consequent extinction of their cheekless rivals, the prosauropod saurischians (Sect. 10.3) of the Triassic.

10.5.1 Fabrosaurids and Heterodontosaurids

The Fabrosauridae, which radiated throughout the world during the Lower Jurassic period, may well have been the ancestral group from which the other ornithischians evolved. Fabrosaurs were lizard-like reptiles that looked superficially somewhat like small theropods. Numerous fossilised bones have been discovered, but only *Lesothosaurus* (Fig. 111) from South Africa is represented by a reasonably complete skeleton. This lightly-built ornithischian predated the Heterodontosauridae (Fig. 88) which were indisputably ornithopods. *Leso-*

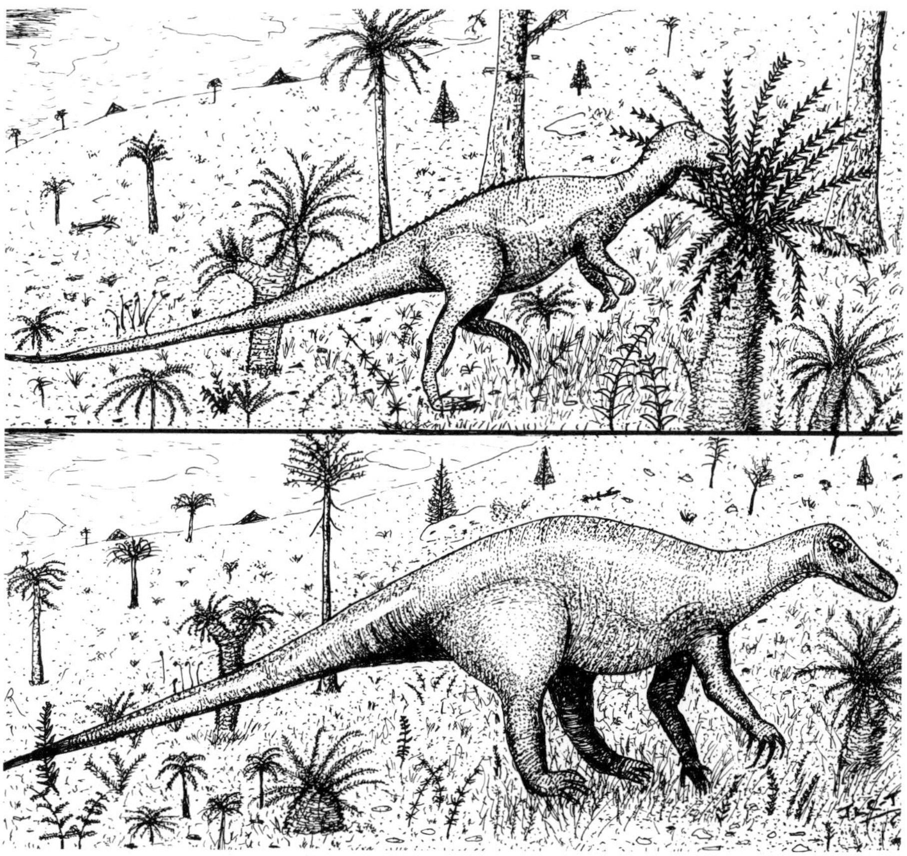

Fig. 111. Lower Cretaceous hypsilophodonts. *Above Hypsilophodon* (length ca. 1.5 m). *Below Tenontosaurus* (length ca. 7.3 m). (Not to scale)

thosaurus had strong arms which could have been used to gather vegetation and pull branches down into its mouth where well-spaced, serrated teeth chopped the tough leaves. Like other fabrosaurs, it lacked the cheek pouches of the later ornithopods. The Heterodontosauridae ('varied-toothed lizards') had differentiated teeth – 2 incisors, 1 canine and 12 molars – on each side of the jaws. The lower canines fitted into deep notches in the upper jaws while the upper canines of the males formed tusks. *Heterodontosaurus* was able to make sideways chewing movements of its lower jaw – a step towards complete mouth processing and disappearance of the need to use gastroliths.

The Fabrosauridae and Heterodontosauridae appear to have been well equipped for survival in a hostile world. Some, like *Lesothosaurus*, were swift and agile. They would have been able to escape from predators by flight. In contrast, *Scutellosaurus* was less speedy but partly armoured. Rows of horny knobs covered its back and flanks, while its long tail would have helped to balance the body on the hind legs while the animal defended itself with its powerful arms. These arms were longer than those of the other known fabrosaurids. This also suggests that *Scutellosaurus* may have browsed on all four legs, relying on its armour for defence in the first instance. This dinosaur's short head was equipped with canine teeth like those of the heterodontosaurids. Some skulls have been found without canine tusks and could have belonged to females. The tusks might well have been a secondary sexual character, used both in sexual display and for defence. Perhaps it was the larger males who defended family groups against attack.

In general, it appears that the heterodontosaurids were speedy and agile bipeds, similar to the fabrosaurids that lived contemporaneously with them, but were able to eat a much greater variety of plant food. They chewed this food with special cheek teeth and held it in their mouths with their fleshy cheeks. Their large and powerful hands may have been used for digging up roots and termite nests as well as for pulling down branches. In 1978, Tony Thulborn postulated that both fabrosaurids and heterodontosaurids might have aestivated during the annual dry season, a view subsequently refuted by Jim Hopson. The controversy has been summarised by Norman (1985).

10.5.2 Hypsilophodontids

The Hypsilophodontidae was a family of medium-sized ornithopod dinosaurs, somewhat resembling the fabrosaurids outwardly. They did not appear until about the Middle Jurassic, however, so the similarity between the two taxa must have been due to convergent evolution. One of the best known of these small ornithopods, *Hypsilophodon* (Fig. 111) from Europe and North America, was first described by T.H. Huxley in 1870. Because of the similarity of its shape to that of modern tree kangaroos (*Dendrolagus* spp.), reconstructions of *Hypsilophodon* have for almost a century illustrated this dinosaur perched, bird-like, in a tree. Not until 1974 was the skeleton reassessed and the conclusion reached that there was no evidence to indicate that it was arboreal. On the contrary, it is now known to have been a speedy terrestrial biped. Other well known genera of hypsilophodontids included *Dryosaurus, Othnielia, Parkosaurus* and *Thescelosaurus* (Palmer 1999).

The lifestyle of the hypsilophodonts resembled that of the fabrosaurids, as mentioned above, but the hypsilophodonts had evolved several adaptive anatomical modifications. One of the most important of these was the development of retaining cheeks that prevented food from falling from the sides of the mouth. In addition, the dentition was better adapted for grinding. The upper and lower teeth met (or occluded) as regular rows and did not interlock alternatively as did those of the fabrosaurids. This would have provided a more efficient grinding surface. At the same time, however, the hypsilophodontids retained more digits, a primitive character, than did the fabrosaurids. They had strong arms with five fingers each and powerful hind legs with four toes. The tail was stiff, and used for stabilisation when the animal was running.

Dryosaurus had sharp, rigid cheek teeth but there were no teeth at the front of the jaw. A horny beak at the end of the lower jaw met a toothless pad on the upper jaw. In *Hypsilophodon* there were incisor-like teeth on the upper jaw which closed against a toothless horny beak on the lower, while in *Parkosaurus* which survived until the end of the Cretaceous period, a horny beak replaced all the front teeth. This dinosaur probably foraged close to the ground, nipping off delicate shrubs from among the low-growing vegetation. Except for *Tenontosaurus* (Fig. 111) from North America, all the hypsilophodonts were relatively small, lightly built bipedal dinosaurs. *Tenontosaurus,* in contrast, was very much larger. Over half of its total length was made up of an extremely thick and heavy tail, while its front legs were longer and stouter than those of other hypsilophodontids, and it probably spent most of the time on all fours. The discovery of a skeleton surrounded by those of five *Deinonychus* (Figs. 86, 87) may have been the result of an encounter with a pack of these formidable predators, although it is more likely that the bodies were brought together by chance – perhaps in a flash flood. Like *Parkosaurus* and also from North America, *Thescelosaurus* survived until the end of the Cretaceous. It was larger than most hypsilophodontids and showed many characters resembling those of the iguanodonts (Sect. 10.5.3), as for that matter did the very much larger *Tenontosaurus.* Whereas the latter measured ca. 7.3 m in length, *Thescelosaurus* was only about 3.5 m long. It, too, was a slow-moving creature but the bony scales on its back would have offered some defence against predators. Unlike other hypsilophodonts, which had three or four toes on each hind foot, *Thescelosaurus* had no less than five (Palmer 1999). As Benton (2004) wrote, the hypsilophodontids had typical elongated feet with hoof-like claws and adapted for running. The overall limb proportions of most genera were similar to those of a fast-moving antelope, especially in regard to the very long shin and foot.

10.5.3 Iguanodontids

The Iguanodontidae evolved in the Middle Jurassic period, about 170 mya, and spread throughout the world, reaching a peak in diversity and abundance by the end of the Lower Cretaceous. From then on, they declined in number and ecological importance and finally disappeared at the end of the period.

Whereas most of the hypsilophondonts were adapted for speed, the iguanodontids were bulky and heavily built with thigh bones longer than their shins. The tail was used to counterbalance the body, and the animals probably adopted a quadrupedal gait for much of the time, because three of the five fingers of their hands had hoof-like nails. The 'thumbs' were developed into prominent spikes which would have been used for defence against contemporary predators such as *Megalosaurus* (Sect. 11.4.2). They could also have been used for sexual display or to tear down foliage. *Iguanodon* (Fig. 97) was only the second dinosaur genus to be discovered. It was first described by Gideon Mantell in 1825, but not until 1877, when the skeletons of more than 30 specimens were discovered in a coal mine at Bernissart in Belgium (Colbert 1968), did the true nature of this impressive dinosaur become apparent. Large herds roamed about, browsing on ferns, horsetails, cycads, conifers and other tough foliage. The head of *Iguanodon* had a prominent snout and powerful jaws with cheek teeth that provided a strong and grinding action, because the bones of the upper jaw moved apart when the lower jaw was raised between them (Norman 1985; Palmer 1999).

Other well-known genera include *Camptosaurus* (Upper Jurassic), *Ouranosaurus* (Fig. 101), *Muttaburrasaurus* and *Probactrosaurus* (Lower Cretaceous). The demise of the iguanodontids could well have been linked with the rise of the hadrosaurs (Sect. 10.5.4), although they were not affected by the hypsilophodontids (Sect. 10.5.2). The reason for this difference may be that they did not compete with the latter family although it was contemporaneous. Hypsilophodontids were very much smaller than iguanodontids, and consequently could have fed upon quite different types of plants. Moreover, their jaws and teeth were rather different, as they were adapted to their own special foods.

10.5.4 Hadrosaurids

The Hadrosauridae or duckbilled dinosaurs was the last family of Ornithopoda to evolve. Hadrosaurids first appeared in the Middle Cretaceous and died out at the end of the period, about 30 my later. Although their bodies were all remarkably similar, like those of early iguanodontids, their heads differed a great deal. Figure 96 illustrates some crested types and a variety of other shapes are shown in Fig. 112. In their day, the duckbilled hadrosaurids were the most common and varied group of ornithopods to evolve. This group probably developed in central Asia and spread throughout the northern hemisphere.

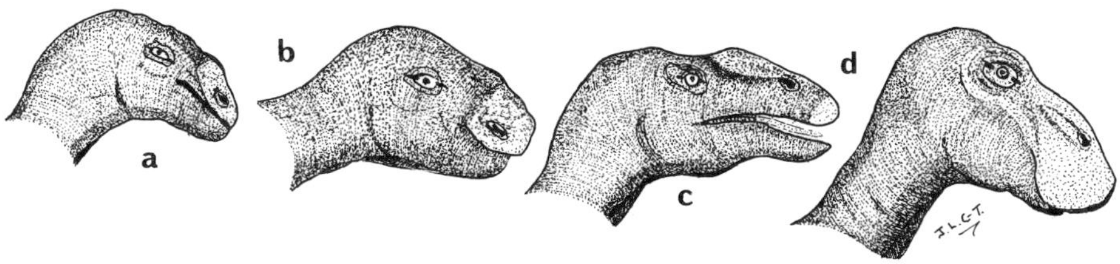

Fig. 112a–d. Crestless hadrosaurid heads (not to scale; cf. Fig. 96). **a** *Bactrosaurus* (total body length ca. 4 m), **b** *Hadrosaurus* (total body length ca. 9 m), **c** *Kritosaurus* (total body length ca. 9 m), **d** *Edmontosaurus* (total body length ca. 13 m; Upper Cretaceous)

Very few genera, apart from the primitive flatheaded *Secernosaurus* from Argentina, found their way into Gondwanaland. This great mass of land had already broken up and, by Upper Cretaceous times, the continents were drifting apart.

The hadrosaurids fed on very tough plant material such as conifer needles and twigs. Like the iguanodontids (Sect. 10.5.3), they had long snouts and heavy jaws with numerous teeth. In the hadrosaurids, these were cemented together by bony tissue to form batteries that ground the food with up and down rather than backward and forward movements. The tough, sharp and horny beaks, from which the common name duckbilled dinosaurs is acquired, would have been well suited to nipping off twigs and stripping leaves from branches. The beaks grew continuously, thus compensating for the wear associated with hard, tough vegetable food. It seems probably that the increase in abundance of hadrosaurids in the Late Cretaceous was associated with the adaptation of their highly efficient jaws to the appearance of flowering plants and disappearance of more primitive types such as seed ferns and bennettitales (Ostrom 1964; Norman 1985).

As reported in the last chapter (Sect. 9.3.3), hadrosaurids nested in large colonies where the young dinosaurs were fed by their parents who returned to the same nest site year after year. The functions of the crests (Fig. 46) and complex narial passages in visual and vocal display have been discussed in Section 9.3.1. The family Hadrosauridae is divided into two subfamilies. In one, the hadrosaurines, the heads were flat and either surmounted by solid, bony crests or there were no crests at all (Fig. 111). In the other subfamily, high domed heads bore flamboyant hollow crests (Fig. 96). These hadrosaurids were known as lambeosaurine duckbills, and were largely confined to North America. The first subfamily, the hadrosaurine duckbills, was the more successful. Distributed throughout North America, Europe and Asia, some of its members were among the last dinosaurs to survive.

The earliest, and one of the smallest hadrosaurids so far discovered, is *Bactrosaurus* (Fig. 112a) from Mongolia and China. This animal had a low, flat head without a crest and a narrow bill like those of the hadrosaurines, but the build of its body was like that of a lambeosaurine duckbill. It may represent an ancestral hadrosaur which had evolved from the iguanodontids. *Bactrosaurus* had typical hadrosaurid dentition capable of grinding tough vegetation. *Hadrosaurus* (Fig. 112b) was the first dinosaur to be discovered in North America – by Joseph Leidy in 1858. Leidy recognised that it was structurally related to *Iguanodon*, (discovered in England and described some 30 years earlier by Gideon Mantell; Sect. 10.5.3). He inferred that *Hadrosaurus* was bipedal. A typical hadrosaurine, it had no crest on its long, low head but, as in *Kritosaurus* (Fig. 112a), there was a large lump on its snout that may have been covered with skin. It has been suggested that this might have been a male sexual recognition character. *Edmontosaurus* (Fig. 112d) was an exceptionally large hadrosaurine duckbill, reaching a length of 13 m. Behind its toothless beak were banks of tightly packed teeth forming 'a veritable grinding pavement in both jaws (Palmer 1999). As these teeth were worn down, they would have been replaced by new ones from beneath. At any one time, there could have been more than 1,000 teeth in the jaws of *Edmontosaurus*. These teeth ground against each other as did those of *Iguanodon* (Sect. 10.5.3). *Edmontosaurus* and its relatives such as *Anatosaurus* would have been able to shred the very toughest plant material.

10.5.5 Psittacosaurids

The last group of bipedal herbivorous dinosaurs to be discussed is the family Psittacosauridae. For many years, the psittacosaurids were thought to be ornithopods and related to *Hypsilophodon* (Fig. 111). *Psittacosaurus* ('parrot lizard'; Fig. 113) from Asia was a small lightly built bipedal dinosaur with a long tail which counterbalanced its body over the hips. Its forelegs were short, and its hands had four fingers with blunt claws, whereas the ceratopsids (Sect. 10.6.2), with which it is believed to have shared a common ancestor, had five. Furthermore, whilst *Psittacosaurus* had no teeth in its beak, the early

Fig. 113. *Psittacosaurus* (Psittacosauridae; Upper Cretaceous; length ca. 2.5 m). (Based on Norman 1985)

ceratopsids (Protoceratopsidae) had teeth in the upper half of the beak. Like the hypsilophodonts (Sect. 10.5.2), the psittacosaurids probably walked quadrupedally for much of the time, but would have been able to run faster on two legs than on four (Palmer 1999). Their parrot-like beaks, with a peculiar rostral bone, suggests that either a new type of plant had evolved that required this type of beak to crop it, or else the development of a parrot-like beak enabled the psittacosaurids and later ceratopsians to feed on plants previously inedible for most dinosaurs (Norman 1985). The appearance of the ceratopsians coincided very roughly with the first appearance of the angiosperms.

10.5.6 Pachycephalosaurians

The Pachycephalosauria had curved, serrated teeth that were ideal for shredding plant material. They also appear to have possessed large eyes and a keen sense of smell. It was originally suggested that their thick, high-domed skulls were adapted for agonistic head-butting (Sect. 9.3.1). This has already been mentioned in the case of *Stegoceras* (Fig. 100). Until recently, most palaeontologists believed that these bone-headed dinosaurs lived in small groups and some, like mountain goats (*Oramnos americanus*) today, engaged in combative head-butting as a means of establishing dominance hierarchies. It is now thought to be more likely that they were used to butt the flanks of opponents (Goodwin and Johnson 1995). Pachycephalosaurs have been divided into two types, high-domed (pachycephalid) and low-domed (homalocephalid) forms. The skulls of the latter were not strongly built, so they most probably used head-to-head pushing contests, whereas the high-domed forms would have indulged in violent contests of flank attacks or head-butting.

10.6 Horned, Plated and Armoured Dinosaurs

The remaining ornithischian dinosaurs were quadrupedal. They included the Ceratopsia or horned dinosaurs, one of whose early families was the Psittacosauridae (the others were Protoceratopsidae and Ceratopsidae), the plated Stegosauria, and the armoured Ankylosauria. *Scelidosaurus* (Fig. 88) from the Lower Jurassic of Dorset, England, was one of the earliest ornithischians. Its body was well armoured with bony plates and spines so that it resembled the archylosaurs in its general appearance (Sect. 9.2.2).

10.6.1 Ceratopsians

Horned dinosaurs have already received considerable attention in these pages (Sect. 9.2.2). They were the last of the ornithischians to evolve before the final extinction of the dinosaurs and were extremely successful and numerous. Ad-

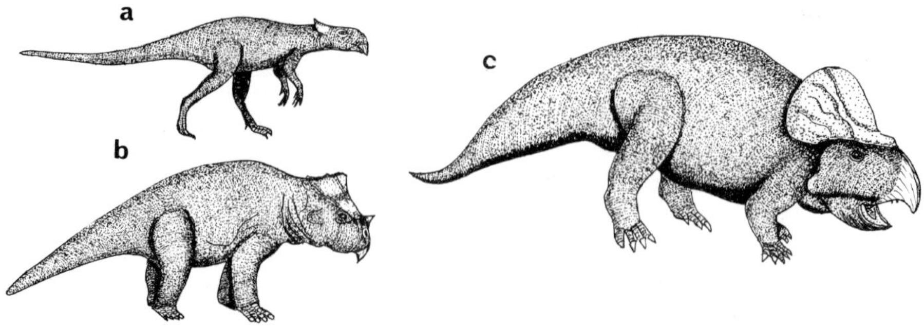

Fig. 114a–c. Protoceratopsids (not to scale). **a** *Microceratops* (length ca. 60 cm), **b** *Bagaceratops* (length ca. 1 m), **c** *Protoceratops* (length ca. 3 m; Upper Cretaceous). (After Palmer 1999)

vanced forms had massive heads armed with a sharp parrot-like beak at the front of the jaw, a varying number of horns and great frills or sheets of bone growing from the back of the skull. These not only protected the neck and shoulders but would have engendered impressive aposematic displays. The success of the ceratopsians was almost certainly related to the ability to eat very tough vegetation thanks to the possession of efficient teeth and powerful jaws. This food probably included the newly-evolved angiosperms that thrived in Late Cretaceous times (Palmer 1999).

The family Protoceratopsidae (Fig. 92) evolved in Asia during the Upper Cretaceous. Only some genera had horns but, like the Psittacosauridae, many of them could walk upright even if they spent most of their time browsing on all fours. *Microceratops* (Fig. 114a) from China and Mongolia is the smallest ceratopsian so far discovered. A lightly built animal, its lifestyle probably resembled that of the hypsilophodonts (Sect. 10.5.2). *Leptoceratops* (length ca. 1 m) from North America was more solid but still essentially bipedal. The Mongolian *Bagaceratops* (Fig. 114b), although no longer than *Leptoceratops*, was even more heavily built and clearly quadupedal. *Protoceratops* (Fig. 114c) must also have spent nearly all the time on all fours, but its hind limbs were still longer than those in front so it was able to rear up and probably run bipedally. A prominent bump on its snout could have been used for ritual head-butting encounters between rival males. The North American *Montanaceratops* (length ca. 3 m) was generally similar in appearance, but had a definite horn on its snout. In addition, it was endowed with an unusually flexible tail, which might have been brightly coloured and used as a sexual signal during the mating reason (Palmer 1999). The Psittacosauridae were discussed in Section 9.2.2.

The great horned Ceratopsidae (Figs. 72, 93–95, 99), the third family of Ceratopsia (Benton 2004) included the most abundant herbivores of the Late Cretaceous in North America. Their long, sharp horns protruded from massive beaks with large bony frills. With their heavy build and pillar-like legs, they were without doubt exclusively quadrupedal. The number of horns varied from a single one on the nose to three in *Triceratops* (Fig. 72) and five in *Torosaurus* (Fig. 91), *Anchiceratops* (Fig. 93), *Chasmosaurus* (Fig. 95b) and *Pentaceratops* (Fig. 95a). Peter Dodson (1996) discussed the number of horns in ceratopsids in detail from a taxonomic viewpoint, simplifying the number of species in the genus *Centrosaurus*, suggesting that *Monoclonius* might actually be the female of *Styracosaurus*, and so on (see also Lambert 1992). Dodson, however, did not discuss any possible ecological significance in the number. It seems to be generally accepted that the variety of ceratopsid horns reflected special epigamic display functions – in addition to the deterrence of predators (Figs. 93, 94: Sect. 9.2.1). Small-horned types would have tended to sway their heads sideways towards the flanks of their opponents, and to display them laterally. Genera with larger horns may have used them for head-to-head ramming, or in complicated pushing and wrestling trials of strength (Fig. 99; Sect. 9.3.1).

Farlow and Dodson (1975) had already concluded that the protoceratopsids would have behaved like small-horned antelopes and delivered sideways blows to the flanks of their opponents. Moderately large frills may have been used as display signals – the larger the frill, the more dominant the animal in its social group. Short-frilled ceratopsids with large, unpaired nose horns behaved more like rhinoceroses. Their formidable weapons might have been employed largely for bluff in agonistic situations. Possibly they were rather solitary animals and did not often need to defend their territories. Finally, the long-frilled genera would have been able to produce an impressive frontal display merely by nodding their heads and swaying them from side to side. When agonistic combat did occur, the brow horns would have been locked together in pushing and wrestling combats (Fig. 99).

Some years ago, the suggestion was made that the ceratopsians might have consumed fleshy fruits that did not require much chewing. J.H. Ostrom, however, thought it more probable that they ate vegetation such as palm and cycad fronds, which could not be exploited by reptiles that lacked the necessary dentition and powerful jaws for this purpose. There is no evidence to suggest that they used their horns to knock down small trees before shearing off and swallowing the leaves and twigs (Dodson 1996).

10.6.2 Stegosaurs

The Stegosauria was a group of moderately large quadrupedal ornithischians that arose in Middle Jurassic times. The first fossil to be discovered was an incomplete skeleton, described by Richard Owen in 1875. Shortly afterwards, numerous specimens of *Stegosaurus* (Fig. 71) were excavated from two quarries in North America. Later, stegosaurs were found throughout the northern continents, even extending into China. They were characterised by a double row of tall spikes running down their backs. The possible functions of these have already been discussed (Sects. 7.5.2, 9.2.2) – they may have had a thermoregulatory or a defensive function, or both. The number of spikes varied between species. The skulls of these dinosaurs were extremely small in relation to the sizes of their bodies, as were the brains within them. That of *Stegosaurus* was about the size of a walnut although the animal's body length was up to ca. 9 m. The ratio of brain mass to estimated body mass (encephalisation quotient) is, however, not related to intelligence mainly because it does not take into account increases in the surface area of the brain due to folding.

Like the iguanodontids and hadrosaurids, the stegosaurs probably roamed the countryside in large herds, but they were exclusively quadrupedal. Nevertheless, the massive hind limbs were over twice the length of the forelegs so that the body sloped forwards from its highest point above the hips. The low position of the head was probably an adaptation for feeding on low ground cover – cycads, ferns and so on. At the same time, stegosaurs would undoubtedly have been able to rear up on their hind legs had they wished. Their feeble jaws and teeth merely cropped large amounts of vegetation, which passed into a vast stomach where it fermented at the posterior end of the gut processor continuum (Sects. 10.1, 10.2). Defence against predators consisted of the spines on the body and a massive tail (Sect. 9.2.2).

10.6.3 Ankylosaurs

The remaining major group of armoured dinosaurs, the ankylosaurs, consisted of two families – Nodosauridae and Ankylosauridae. Ankylosaurs were characterised by a great variety of spikes and armour plating. Some genera possessed cranial horns, and the ankylosaurids were endowed with massive tail clubs (Fig. 90), which were probably used both in defence against predators and to combat rivals. In addition, the broad, flattened and armoured skulls might have been employed in head-to-head pushing (Sampson 1997). In contrast, the earlier and more primitive nodosaurids had narrower skulls and lacked tail clubs, but many of them had enlarged spikes which might have interlocked with those of opponents in dominance contests (Coombs 1990). Their bodies were also covered with bony plates from head to tail. Well-known genera include the Lower Cretaceous *Hylaeosaurus*, whose remains were described by Gideon Mantell in 1832, and its contemporary *Polacanthus* (Fig. 115), also from southern England. Because the fossils are incomplete, the arrangement of the spines is not known and it is possible that the two animals are the same. *Sauropelta* and *Silvisaurus* date from the Lower Cretaceous of North America, *Nodosaurus* (Fig. 115) and *Panoplosaurus* (Fig. 115) from the Upper. The diminutive *Struthiosaurus* (length ca. 2 m) might well have been an island form. Many large animals have evolved dwarf species when confined to islands where food supplies are limited. In contrast, *Sauropelta* was ca. 7.6 m long and weighed about 3 tonnes (Palmer 1999).

In addition to their formidable spines, the ankylosaurs were encased in thick, defensive, body armour. Rows of horny plates with raised keels ran from the head to the tip of the tail in *Sauropelta*. *Silvisaurus* was covered with thick bony

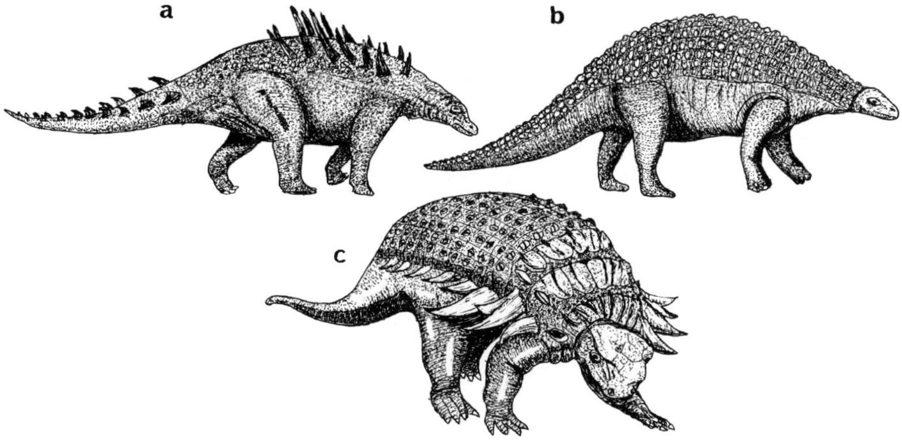

Fig. 115a–c. Nodosaurids. **a** *Polacanthus* (Lower Cretaceous; length ca. 4 m), **b** *Nodosaurus* (Upper Cretaceous; length ca. 5.5 m), **c** *Panoplosaurus* (Upper Cretaceous; length ca. 10 m; after Palmer 1999)

plates, while the armour of *Nodosaurus* consisted of narrow rectangular plates alternating with wider plates studded with bony nodes. Whereas most genera probably squatted, retracted their legs, and relied upon their armour for defence against predators, *Panoplosaurus* (Fig. 115), although well armoured, could have charged and struck the enemy with its large shoulder spikes. Its forelegs were particularly strongly built and well endowed with muscles, suggesting that this dinosaur was quite manoeuverable. The shoulders and hips of the ankylosaurs were well modified to bear the great weight of armour, while the legs were strong, with broad hoofed feet (Palmer 1999).

Ankylosaurid armour and spines were even heavier and more formidable than those of the nodosaurids. *Dyoplosaurus* (Fig. 89) from the Upper Cretaceous of Canada had a toothless beak, small cheek teeth, an armour-plated body, and a tail ending in a bony club remarkably like that of *Euoplocephalus* (Fig. 90; Sect. 9.2.2). Indeed, it is quite possible that once again two sets of fossils belong to the same species!

The amount of bone was much reduced in most nodosaurid skulls, but the head armour of the ankylosaurids was extensively developed. In *Talarurus* (Fig. 116) from Mongolia, the skull broadened out from the toothless beak and small cheek teeth of the mouth to a pair of hefty, bony spikes at the rear. Another pair of spikes projected from the side of the head. The massive head of *Saicharnia*, also from the Upper Cretaceous of Mongolia, was armoured with bony nodules, while the skull of *Euoplocephalus* from North America had a skull wider than any other part of its body. Its snout was blunt and, as usual, the front of the mouth consisted of a toothless beak. Massive armour, pointed spines, and a tail ending in a great bony club, were the main distinguishing features of the Ankylosauridae.

Fig. 116. Ankylosaurids. *Above Talarurus* (Upper Cretaceous; length ca. 5 m). *Below Ankylosaurus* (Upper Cretaceous; length ca. 10 m). (After Palmer 1999)

Many of the nodosaurids had toothless beaks, but *Silvisaurus* retained teeth in its upper jaws. Ankylosaur teeth were small in relation to the size of the skull. They were laterally compressed and leaf-shaped. Those of nodosaurids were slightly larger, both absolutely and relative to skull size. Whereas the ankylosaurids would have cropped low-level vegetation indiscriminately, the nodosaurids, with their narrow snouts, may have been more selective in their choice of food.

A complex of chambers and air passages was characteristic of both ankylosaurids and nodosaurids. Like the crests of hadrosaurs (Sect. 9.3.1), these might have been used to produce sounds for social communication and agonistic threat. The nasal passages of the nodosaurids were relatively simple paired tubes leading from the nostrils to the back of the throat, but those of the ankylosaurids followed an S-shaped course through the head. On either side, there were additional branches or sinuses that may have had several functions in addition to sound production. They might have improved the sense of smell, or served to filter, warm, and moisten inspired air. In 1977, Teresa Maryanska discovered bones within the nasal passages of *Saichania* and *Pinacosaurus* that resembled the turbinal bones in the noses of mammals. Covered by moist membranes, these bones would have had just those regulatory functions (Norman 1985).

10.7 Conclusions

The herbivorous adaptations of Late Triassic and Early Jurassic dinosaurs have been outlined by Galton (1986). The Jurassic period was marked by an abundance of primitive plants and large sauropod dinosaurs. During the Cretaceous, however, flowering plants evolved and radiated, while some of the more primitive types of plant became extinct. It seems highly probable – although impossible to prove – that there was a causal connection between these events, and that the success of the ornithopod dinosaurs was a reflection of their ability to feed upon the newly evolved phanerogams (see discussion in Tiffney 1997).

Indirect evidence as to what herbivorous dinosaurs fed on is plentiful. This is based on contemporary vegetational surveys, and on hypotheses regarding foraging abilities inferred from functional morphology. There is little direct evidence, however, to indicate what plants they actually ate. Fossil trackways sometimes provide information (Farlow and Chapman 1997; Lockley 1997). For instance, the tracks of Cretaceous dinosaurs around fossil tree trunks, described by L.R. Parker and R.L. Rowley Jr. in 1989, suggest the shuffling footsteps of browsing hadrosaurs. Seeds, conifer twigs and pine needles have been found in the body cavity of a fossil *Edmontosaurus*. These could have been the stomach contents or, equally well, represent debris washed into the decayed carcass. Finally, the coprolites of herbivorous dinosaurs – although much less common than those of carnivores – have occasionally been found. Even so, they have often been partly decomposed, or devoured by dung beetles before fossilisation and it is difficult to know what species of dinosaur produced them. Most are composed mainly of conifer stem tissue, indicating a highly fibrous diet. Many have been found in close association with the hadrosaur *Maiasaura*, which had a battery of grinding cheek teeth that could easily have chewed wood (Chin 1997).

11 Carnivorous Dinosaurs

11.1 Introduction

The saurischian dinosaurs are classified into two suborders, the Sauropodomorpha (Chap.10) and the Theropoda. With few exceptions, the latter consisted of bipedal, flesh-eating Saurischia. It was traditionally divided into the infra-orders Ceratosauria and Carnosauria. The first of these consisted of small light-weight hunters and medium-sized carnivores with a slashing or killing claw (Sect. 9.2.1), while the largest predators comprised the Carnosauria, which contained the families Megalosauridae, Spinosauridae, Allosauridae and Tyrannosauridae. These are now recognised as having little in common, apart from their size (Benton 2004). The term carnosaur is nevertheless a convenient one to use provided that this limitation is recognised.

The Herrerasauridae and other early or primitive carnivorous dinosaurs (Sect. 9.1) will not be discussed further in the present chapter. The ceratosaurs, following Palmer (1999), are subdivided into the families Ceratosauridae and (division Maniraptora) Coeluridae, Compsognathidae, Ornithomimidae, Oviraptoridae, Dromaeosauridae and Troodontidae. Mention will also be made of the Therizinosauridae or segnosaurs and the carnosaurs.

The Theropoda comprised a group of carnivorous dinosaurs of bewildering variety. Their relationships are far from clear because, although many isolated fossil bones and teeth have been found, few reasonably complete skeletons or skulls have been discovered. Although this book is not concerned with the problems of taxonomy – which have been considerably simplified – it would be a mistake to think that types grouped together on account of shared predatory habits and associated skeletal form are necessarily closely related.

11.2 Dentition

Over the years, there has been much speculation as to what the prosauropods actually fed on (Sect. 10.2). The subject has been reviewed by VanHeerden (1997) who presented a table (see Table 4) adapted from the work of P.M. Galton contrasting the features of the jaws and teeth of carnivorous and herbivorous reptiles past and present. When the articulation of the jaw is in line with the rows of teeth, as in carnivorous reptiles, the jaws close in a scissors-like manner. In con-

Table 4. Features of the jaws and teeth of herbivores and carnivorous reptiles. (Based on VanHeerden 1997)

Herbivores	Carnivores
Jaw articulation below the rows of teeth	Jaw articulation in line with the rows of teeth
Teeth closely spaced	Teeth with large gaps between them
Tooth crowns twisted so that they overlap in side view	Tooth crowns in line along middle of the jaw
Teeth leaf-shaped	Teeth taper from root to tip
Teeth have coarse serrations at about 45° to the edge of the crown that occur more towards the tips	Teeth have fine serrations at right angles to the edge of the crown
Teeth are more or less the same size (because there is less breakage)	Teeth are of different sizes. New teeth replace broken and lost ones

trast, when the jaw articulation is offset vertically from the rows of teeth, the force of the bite is distributed over a greater distance and less force is applied at any one point. Widely spaced teeth are better for piercing; closely spaced teeth form a cutting edge. When teeth are set in line along the middle of the jaw, they are less likely to be broken by struggling prey than when their crowns are twisted. Moreover, leaf-shaped herbivorous teeth are weaker than are spiky, tapering teeth – which work like daggers and can easily be extracted from the body of the prey when the jaws are opened. Teeth with fine serrations can cut harder material, such as bone, than can teeth with coarse serrations.

Despite the fact that plant material tends to be more resistant than animal flesh, it is argued that, in general, herbivorous dinosaurs would have selected softer vegetation and avoided the woody parts. From this, it is clear that most, if not all, of the basal sauropodomorphs would have been poorly adapted to a carnivorous diet – as would all the herbivorous forms discussed in Chapter 10. The theropods were probably all more or less carnivorous, even the toothless forms described below (Sect. 11.3.1).

11.3 Ceratosaurs

Whereas the Herrerasauridae (Sect. 9.1) are known only from the Late Triassic of South America, the Ceratosauria was a much larger group that persisted from the Late Triassic to the Upper Cretaceous period. Possibly, the oldest ceratosaur genus known so far is *Coelophysis* (Fig. 117; family Podokesauridae) of which many well-preserved specimens have been found in Connecticut and New Mexico. They are of all ages – from newly hatched babies to full-grown adults. However, the latter only measured ca. 3 m in length and about 1 m in height, which is small for a dinosaur. Built for speed with light, hollow bones, they probably weighed little more than 20 kg. They were highly active

Fig. 117. *Coelophysis* (Podokesauridae; Upper Triassic; length ca. 3 m)

predators with pointed heads and sharp, serrated teeth. Their long, slender hind legs had bird-like feet and three toes, each with sharp claws. There were four fingers on the hands, but only three of these were strong enough to grasp prey (Charig 1979). These ceratosaurs probably chased almost anything that moved, including the early shrew-like mammals, as they roamed the upland forests hunting in packs close to streams and lakes. Two of the adult skeletons from New Mexico contained the bones of tiny *Coelophysis* in their bodies. This was initially thought to indicate that the animals were ovoviviparous, but their hip bones were too narrow for this to have been the case. The only explanation is that these dinosaurs were cannibalistic. Since cannibalism is widespread among animals of all kinds, it is not surprising that it should have been practised by at least some theropods (Currie 1997a).

Another member of the Podokesauridae, *Procompsognathus* (length ca. 1.2 m) was a rapacious little theropod which inhabited the deserts of northern Europe during the Upper Triassic. It would have chased its prey, consisting of insects, lizards and the like, on its long legs. It ran with only three of its four toes touching the ground. Each hand had five fingers. This is a primitive feature, since the evolutionary trend was toward fewer fingers and toes (Palmer 1999). Its diminutive relative *Saltopus* was only ca. 60 cm in length and probably weighed as little at 1 kg. *Saltopus* still retained five digits on each hand, although the fourth and fifth were reduced. It was more advanced, too, in its skeletal structure. Only a single specimen has been found, however, and it is possible that this animal was not a dinosaur but a non-dinosurian archosaur (see Benton 2004).

11.3.1 Maniraptora

The remaining families of ceratosaurs (Division Maniraptora) shared a number of characteristics with birds. As the name suggests, these are especially evident in the fore limbs, and the class Aves is actually included within the Maniraptora (Sect. 11.5). The family Coeluridae flourished from the Upper Jurassic to the Lower Cretaceous. *Coelurus* ('hollow tail') from the Upper Jurassic (length ca. 2 m) was an active, highly cursorial predator that inhabited the forests and swamps of North America. Its long arms and hands with three clawed fingers could have grasped small animal prey and the eggs of other dinosaurs.

Compsognathus (Fig. 118; family Compsognathidae) from Europe was a contemporary of the coelurids and resembled them closely. It had typical hollow bones, a long neck and tail, two forceps-like fingers on its short arms, and long, slender legs with three long clawed toes pointing forwards and a small fourth toe projecting backwards. Measuring only ca. 60–90 cm in length – not much larger than a chicken – it probably weighed only about 3.6 kg. Its long skull was unusually large in relation to the size of its body. The first specimen to be discovered was found in a lithographic limestone quarry in Germany, in 1861. The fossil was very well preserved in fine-grained sediment. *Compsognathus* may have lived on insects, scorpions and small reptiles as well as vegetable matter. Part of the skeleton of a lizard (*Bavarisaurus*) can be seen in the gut region of the

Fig. 118. *Compsognathus* (Compsognathidae; Upper Triassic; length ca. 60 cm)

holotype of *Compsognathus longipes* (Chin 1997). Judging by the proportions of the limbs and its long tail, John Ostrom in 1978 deduced that *Bavarisaurus* must have been an agile and speedy ground-living lizard. In order to catch prey such as this, *Compsognathus* would of necessity have possessed keen sight, speed, manoeuverability and rapid reactions (Norman 1985).

Sinosauropteryx (Fig. 126a; Compsognathidae), one of the first of the feathered dinosaurs to be discovered, was found in the same region of China as *Protarchaeopteryx*, which was also a 'dino-bird' (Sect. 11.5). Although very similar to *Compsognathus, Sinosauropteryx* was slightly larger (ca. 1.2 m long) and an equally effective predator. Fossilised stomach contents included a lizard and a small mammal.

The Oviraptoridae, a small family of toothless therapods, lived in eastern Asia during the Upper Cretaceous. *Oviraptor* (Fig. 104) was a typical maniraptoran ceratosaurian. As we have seen (Sect. 9.3.3), it has probably been misnamed and could even have been a herbivore (Sect. 10.1). Its short parrot-like beak and powerful jaws might have been adapted for shearing fibrous plant tissues, as well as cracking thick-shelled dinosaur eggs (Norman 1991) and crushing the shells of molluscs. However, Currie (1997a) argued that although they were clearly adapted to produce powerful bites, the beaks were hollow and filled with air. They would not have been suitable for crushing thick molluscan shells and, in any case, oviraptorids were most common in semi-arid and arid environments. The sharper claws and body proportions of *Oviraptor* suggest that it would have preyed on smaller animals such as lizards and mammals, these to be swallowed whole. The most likely possibility seems to be that, as with members of the ostrich-like Ornithomimidae, *Oviraptor* was omnivorous, feeding upon anything, animal or vegetable small enough to be swallowed without being chewed.

The Ornithomimidae was a specialised dinosaur family that evolved from the coelurosaurs at the end of the Jurassic period. Their toothless jaws were long and slender – quite unable to produce enough power to tear carcasses apart or kill anything but the smallest animals. It has long been assumed that they were omnivorous and fed on invertebrates, small vertebrates, eggs and easily digested vegetable matter, including fruit, berries and seeds. The best-known genera are *Ornithomimus* and *Struthiomimus* from the Upper Cretaceous of North America, both of which were ca. 3.5 m in length. *Ornithomimus* inhabited dense cypress swamps and forests. It would have sprinted with its body parallel to the ground and balanced by its long, stiff, outstretched tail. Its neck would have curved upward in an S-shape – as would the necks of most bipedal dinosaurs – so that it could see ahead, while its arms dangled in front ready to grasp any potential food item. *Struthiomimus* existed slightly earlier than *Ornithomimus*. Its arms were longer and it probably hunted along river banks in more open country (Palmer 1999).

The largest of the ornithomimids was *Gallimimus* (Fig. 119) from Mongolia, which measured ca. 4 m in length. All the 'ostrich' dinosaurs appear to have been extremely fast, with big eyes and large brains. Those of *Dromiceiomimus*

Fig. 119. *Gallimimus* (Ornithomimidae; Upper Cretaceous; length ca. 4 m)

were proportionately larger than the brains of any modern land mammals other than primates. There is still no consensus as to the feeding habits of the ornithomimids. Some authors consider them to have been predators, others have argued that they might have been herbivorous. I think this to have been far less likely. Barsbold and Osmólska (1990) considered that some of them were moderately fast, wading animals, moving about on spongy substrates and feeding upon fishes and molluscs. *Dromiceiomimus* probably hunted after dark, chasing lizards and small mammals through the deciduous woodlands in which it lived (Palmer 1999). In a review of the theropods, Currie (1997b) suggested that the ornithomimids were the fastest of all the dinosaurs. Like ostriches (*Struthio camelus*), they could probably run at speeds up to 80 km/h, according to Currie (1997a), though this figure has been disputed.

11.3.2 Dromaeosaurids

The family Dromaeosauridae comprised some of the most ferocious predatory dinosaurs of the Cretaceous period. Not only did they have the lightweight bodies and speed of the early coelurosaurs, from which they probably evolved, but they possessed much larger brains and huge sickle-shaped killing claws on the second toe of each hind leg (Sect. 9.12). As early as 1964, J.H. Ostrom suggested that they might well have hunted in packs, rather as Cape hunting dogs

(*Lycaon pictus*) do today. *Deinonychus* (Figs. 86, 87) and *Velociraptor* (Fig. 92) have already been mentioned in this context.

When dromaeosaurs were running, their killing claws were flicked upwards and held clear of the ground while the body was poised horizontally and balanced by the outstretched tail. This was kept rigid by bundles of bony rods growing from the vertebrae. It is essential to maintain balance when standing on one leg and striking the prey with the other, and this is assisted by having a stiffened tail (Sect. 7.3).

The discovery in Manitoba of five complete skeletons of *Deinonychus* beside that of a *Tenontosaurus* (Fig. 111) may well be those of a pack of these predators, killed by some chance event whilst attacking the large herbivore. Alternatively, of course, the assemblage of bodies might have been accidentally brought together after death, perhaps washed down by a flood into a hollow or river basin (Palmer 1999). The possible predatory attack has been superbly reconstructed in a drawing by John Sibbick in Norman (1985). Sibbick also illustrated, in a sequence of drawings, how an individual *Deinonychus* might have hunted fleet-footed prey, such as *Hypsilophodon* (Fig. 111), by leaping onto their backs, subduing them with arms and teeth and administering the *coup de grâce* with the killing claws.

11.3.3
Troodontids

Although smaller and lighter in build than dromaeosaurs, the Troodontidae (= Sauronithoididae) were also fast, intelligent predators. *Sauronithoides* (Fig. 120) from Mongolia is credited with having been one of the most intelligent of the dinosaurs. (Another is *Stenonychosaurus*, a member of the same

Fig. 120. *Sauronithoides* (Troodontidae; Upper Cretaceous; length ca. 2 m). (After Cloudsley-Thompson 1977 after Russell 1971)

family.) The bird-like skull of the troodontids, long, low and light in weight, contained a relatively enormous brain. *Sauronithoides* evidently had fast reflexes and well-developed senses. Its eyes were huge (ca. 15 mm diameter) judging from the size of the sockets. This indicates that the animal was probably a crepuscular or nocturnal predator, as were the other members of the family. Moreover, the position of the eyes on the skull suggests the possession of binocular vision (Sect. 9.3.1) and all that this implies. The troodontids had sharp teeth, serrated on the posterior edges, and a slightly enlarged claw on the second toe, which was normally folded back. This would have been used in defence, and also to slash potential prey such as lizards and small mammals.

11.3.4 Therizinosaurids – Segnosaurs

The known remains of these enigmatic dinosaurs from Mongolia and China consist of the incomplete skeletons of at least four genera. The therizinosaurids ('scythe lizards') or segnosaurs were most unusual theropods, ca. 3–7 m long with a massive pelvis and short tail. They had long hind legs with four-toed feet while the arms, which were shorter, bore hands with three elongated fingers and enormous sickle-like claws. The best fossil of *Therizinosaurus* consists of a huge arm, 2.5 m in length with claws 0.7 m long. This does not include the horny part of the claw, which would have made it even longer. The neck was powerful, but the skull was small, with weak, peg-shaped cheek teeth and a toothless beak in front.

Fig. 121. *Alxasaurus* (Therizinosauridae; Upper Cretaceous; length ca. 7 m)

Because many of the fossils were associated with aquatic environments, the segnosaurs were at one time thought to have been piscivorous. The suggestion has also been made that the claws were formidable weapons for slashing through the skin of their prey, or even for tearing termitaria apart like the claws of anteaters (Myrmecophagidae). The discovery of *Alxasaurus* (Fig. 121) in China, however, provided information suggesting that the therizinosaurids were more probably leaf eaters like giant ground sloths (*Megatherium*). Perhaps they sat, balanced tripedally on their rear limbs and tail, raking tree branches with their claws and passing them to their toothless beaks (Maryańska 1997). This is how they are illustrated in Fig. 121 (see Blount and Crowley 2001).

11.4 Carnosaurs

The largest terrestrial carnivores of all time, the carnosaurs, formed a group of theropods that, although not closely related (Sect. 11.1), show considerable resemblance to one another, probably as a result of convergence. The families Megalosauridae and Allosauridae were, however, more nearly related to one another than to the Tyrannosauridae. They had neck vertebrae with a ball and socket joint at the front and a socket on the back, whereas the neck vertebrae of the Tyrannosauridae were almost flat at both ends. All the carnosaurs had well developed olfactory and optic lobes of their brains. One or two genera may well even have possessed stereoscopic vision (Sect. 9.3.1; Molnar and Farlow 1990). Traditionally, carnosaurs have been illustrated walking with the vertebral column inclined at an angle of at least 45 ° to the horizontal, as in the drawing of *Tyrannosaurus rex* (Fig. 122). This, however, is no longer universally accepted

Fig. 122. *Tyrannosaurus* (Tyrannosauridae; Upper Cretaceous; length ca. 12 m). (Cloudsley-Thompson 1994)

as being correct. The idea was originally based on the work of E.D. Cope in 1886, supplemented by analogy with kangaroos. The centre of mass of the body must lie in the immediate area of the hips. To be balanced, the trunk would either have been held more steeply or its weight was counterbalanced by the tail raised from the ground (Alexander 1985). It is shown in the latter position in most of the illustrations of carnosaurs on the following pages.

11.4.1 Predation and Scavenging

Estimates based on fossil trackways suggest that medium-sized and large theropods walked at speeds of about 5–10 km h^{-1}. This is quite as fast or even faster than large mammals of today. It would suggest that these flesh-eating dinosaurs were nomadic animals and ranged widely in search of food. Although speeds considerably faster than this have been proposed by several authors, including Bakker (1987), it does not seem that greater speeds would have been necessary, because most of the smaller dinosaurs on which they preyed would not have been able to move any faster (Farlow and Chapman 1997).

The relatively small size of the arms of the carnosaurs has often given cause for surprise and speculation as to their function. The extreme is reached in *Tyrannosaurus* (Fig. 122) in which the shoulder bone was relatively large while the limb was tiny and bore a small two-clawed hand. It has been postulated that this hand might have been used for grappling the partner during mating. An alternative suggestion is that the arm served to anchor the front of the body while the gigantic carnivore struggled onto its feet after resting on the ground (Norman 1985; Molnar and Farlow 1990). Quite possibly, both functions were employed. Even so, it is difficult to imagine why the front legs were reduced to such an extent.

When closing with their prey, the carnosaurs would have attacked with open jaws, grabbed the victim and shaken it violently. The shock of this would have caused death, as it does when animals are killed in this way by crocodiles, sharks and killer whales (*Orcinus* spp.). It is highly probable that, like top predators today, the carnosaurs were not only predators but took to scavenging when opportunities occurred (Currie 1997a). That they were also opportunistic cannibals, is proved by the presence of distinctive tooth marks on fossils of *Majungatholus* from Madagascar.

Carnosaurs were considerably larger than the predatory mammals that replaced them. The maximum sizes of predatory dinosaurs were the result of a conflict between two demands: population must be low enough to avoid over-exploitation of food supply, but large enough to prevent chance extinction. Studies of fossil tracks suggest that heavy carnivores with narrow, bird-like feet, did not leave firm sandy surfaces, whereas herbivorous dinosaurs with much broader feet would have been able to browse in swampy areas beyond their reach. This would have helped to maintain the balance between herbivores and carnivores (Lockley 1997).

The stochastic probability of extinction increases as population size decreases ('stochastic' refers to patterns resulting from random effects). Flesh-eating dinosaurs solved the conflict by one or more combinations of the following: (1) They maintained larger population densities and/or faster rates of turnover than might be expected of their herbivorous prey, and probably higher than that of large contemporary mammals. (2) Their oviparity produced a larger number of viable offspring than viviparity would have achieved under similar environmental conditions. (3) They fed on different prey when they were young than when adult. (4) They consumed less food than would be expected for equally large meat-eating mammals. The most probable factors involved in enabling the larger carnosaurs to survive were that their mass-specific food consumption was lower than might be expected while, at the same time, their reproductive potential was higher than that of carnivorous mammals (Farlow 1993).

Valid genera of carnosaurs so far known number nearly 20, while there are almost double that number of dubious genera. The latter are based on fragments of fossilised bones (Norman 1985).

11.4.2 Megalosaurids

The Megalosauaridae was a family of very large, bipedal carnosaurs. Their heads were slightly built, high and narrow, but strengthened by bony ridges and equipped with powerful jaws lined with serrated fangs. Their arms were short and strong, with three large clawed digits, their hind limbs large and powerful. There were three large, clawed toes and a small, backwardly-directed, fourth toe. About 15 genera have been described, some of them rather doubtful, mainly from Europe and Asia and dating from the Early Jurassic to the Late Cretaceous. Megalosaurids may well have seized their prey with their clawed hands and feet, while taking large bites of flesh (Lambert 1992).

11.4.3 Spinosaurids

The Spinosauridae was a specialised group of large Lower to Upper Cretaceous carnosaurs. The crocodile-like snout of *Spinosaurus* (Fig. 101d) from Africa suggests a connection with *Baryonyx* (Fig. 123) from England and possibly Africa, while *Acrocanthosaurus* (Fig. 101c) from the Lower Cretaceous of North America seems to have been allied to the Allosauridae.

The claws of *Baryonyx* (Fig. 123) reached a length of 30 cm but, because no attachment was found to the animal's skeleton, could have belonged to either the hind or the forelimbs. Since the latter were unusually thick and powerful in proportion to those of other theropods, it is usually assumed that the claws were carried on them, the animal probably bipedal. The hands bore three fingers with the claw presumably on the largest, the thumb. The neck was long and fairly straight while the skull was long and narrow like that of a crocodile, and *Bary-*

Fig. 123. *Baryonyx* (Spinosauridae; Lower Cretaceous; length ca. 6 m)

onyx had twice as many teeth as most theropods. It was almost certainly piscivorous as partly digested fish scales lay inside the fossilised rib cage. *Carcharodontosaurus* (Lower Cretaceous) and *Spinosaurus* (Upper Cretaceous) had a similar tooth and jaw shape (Ryan and Vickaryous 1997). *Baryonyx* probably hunted on all fours along riverbanks, hooking fish out of the water with its large claws in the manner of grizzly bears (*Ursus* spp.).

Some of the spinosaurids, such as *Acrocanthosaurus*, which was enormous, had backbones with low spinal crests while others, including *Spinosaurus* had extravagant sails on their backs. These, as we have seen, might have served in thermoregulation or agonistic display (Sect. 9.3.1) – quite possibly in both functions. *Ouranosaurus* (Fig. 101a; Iguanodontidae), a vegetarian contemporary of *Spinosaurus* from the same regions of Africa, also had a large sail on its back. This may well implicate the influence of some climatic or environmental factor in the evolution of these structures (Norman 1985; Palmer 1999). *Spinosaurus* was one of the largest carnivorous dinosaurs to evolve, so the sail, although it must have been vulnerable, was clearly not a disadvantage. The crocodile-like shape of its jaws and the straight, rather than curved teeth suggest that, like *Baryonyx* and the gigantic tyrannosaurid *Carcharodontosaurus*, *Spinosaurus* may actually have been piscivorous (Ryan and Vickaryous 1997). This suggestion is strengthened by the fact that its arms were larger than is usual among carnosaurs, and it might have spent much of the time on all-fours prowling the banks of rivers in search of large fishes and other inshore aquatic vertebrates.

11.4.4 Allosaurids

The largest carnosaurs of the Upper Jurassic were members of the family Allosauridae. Similar in build to the megalosaurids (Sect. 11.4.2) but even larger, they colonised all the land surface of the world. *Allosaurus* (Fig. 124) is the largest and best-known genus. About 12 m in length, it must have been at least 4.6 m high and weighed some 1–2 tonnes. There were strong bony prominences above the eyes, and a long, narrow ridge ran from between these down to the tip of the snout. The massive skull was lightened by the presence of several large fenestrae, and the bones were only articulated loosely, as in several other carnosaurs. These features would have conferred flexibility and added to the strength of the skull. The suggestion has been made that the teeth of *Allosaurus* might have functioned like the canine teeth of sabre-tooth tigers, and that these dinosaurs slashed their prey and then waited for it to succumb to its wounds. Skull bones with movable joints could have allowed the jaws of *Allosaurus* first to gape open, then move apart to bite huge mouthfuls from its victims (Blount and Crowley 2001).

Some palaeontologists believe that *Allosaurus* would have been too heavy and clumsy to have caught living prey and that it probably fed on carrion (Sect. 11.4.1). Others are of the opinion that it was quite agile for its size and may even have hunted in packs, like prides of lions, to overwhelm the giant plant-eating dinosaurs of its time. These would have included *Apatosaurus* and

Fig. 124. *Allosaurus* (Allosauridae; Upper Jurassic – Lower Cretaceous; length ca. 12 m)

Diplodocus, as well as a number of smaller forms such as *Camarasaurus*, *Ceratosaurus*, *Stegosaurus* and *Camptosaurus* (Norman 1991). Bones of *Apatosaurus* have, indeed, been found in North America bearing the marks of teeth resembling those of *Allosaurus*. I believe that, like the larger mammalian carnivores today, *Allosaurus* would have been an opportunist, preying on anything it could catch and scavenging on any carrion it happened to find (Sect. 11.4.1).

The remains of at least 44 fossilised *Allosaurus* skeletons have been found in one bone bed in Utah. This accumulation of bones appears to have taken place over a relatively long period of time and provides no evidence of gregarious behaviour. It is more likely that the site was a 'predator trap'. In predator traps, herbivores become mired in quicksand, tar or mud. Their cries and struggles attract the attention of carnivores which themselves become trapped. Their carcasses then attract more predators and scavengers, which are also trapped, attracting still more carnivores to their doom. The final result is that far more predators are trapped than prey (Currie 1997b). Other explanations for the mainly monospecific occurrences of *Allosaurus*, as well as of *Coelophysis* (Fig. 117; Podokesauridae), in such bone beds have also been proposed. Farlow (1987) suggested that they might represent habitat preferences and that the animals died during a particular stage of their reproductive cycle. This could explain the presence of a variety of ontogenetic stages. Coombs (1990) thought that a catastrophic event might have driven the animals together. A group of two juveniles as well as a subadult and an adult *Troodon* (Sect. 11.2.3) may represent a family unit that perished together on the shores of a freshwater lake (Horner 1997).

Allosaurus may have been a ponderous predator, but the smaller *Neovenator* (ca. 9 m) from the Middle Cretaceous of England would easily have outrun its prey – probably *Iguanodon*, the remains of which have been found in the same site. Through modelling the mechanics of the skull of *Allosaurus* by computer, David Norman (1985) revealed that it was many times stronger than would have been necessary merely to remove flesh. He therefore concluded that these carnosaurs were lone ambush predators who rushed at their unsuspecting prey, took a huge bite and then withdrew until the unfortunate victim died from shock and loss of blood, as suggested above.

11.4.5 Tyrannosaurids

No dinosaur has captured public attention so much as the awesome *Tyrannosaurus rex* (Figs. 85c, 94, 122) from North America. Coprolites (fossilised faeces) containing the shattered bones of herbivorous dinosaurs suggest that *Tyrannosaurus* crushed food in its powerful jaws before swallowing it. Less well known is *Gigantosaurus* (Fig. 125), also of the family Tyrannosauridae, from the Upper Cretaceous of Patagonia, and significantly larger. *Carcharodontosaurus* from North Africa was another huge bipedal flesh-eater (length ca. 8 m), but it was lightly built with a large head and shark-like fangs. Its relationships, however, are uncertain and it could possibly have been related to the

Fig. 125. *Gigantosaurus* (Tyrannosauridae; Upper Cretaceous; length ca. 14 m) attacking a hadrosaurid (length ca. 9 m)

Compsognathidae (Sect. 11.3.1) rather than to the Tyrannosauridae. Unlike *Iguanodon*, however, which walked on all fours, the giant carnosaurs would have walked and run bipedally, their bodies almost horizontal and balanced by the heavy tail (Figs. 124, 125). Another gigantic Upper Cretaceous tyrannosaurid was *Tarbosaurus* from Mongolia (length ca. 14 m). It could have preyed on the hadrosaurs and armoured dinosaurs that lived in the same environment. It was somewhat more lightly built than *Tyrannosaurus*, but had a longer skull.

Smaller tyrannosaurids included *Albertosaurus* (length ca. 8 m) and *Daspletosaurus* (length ca. 8.5 m) from North America, and *Alioramus* (length ca. 6 m) from Mongolia. *Albertosaurus* differed from the other tyrannosaurids in being much more lightly built, but its lifestyle was probably very similar. The different genera must have evolved in different places or at different times during the Upper Cretaceous – which lasted for nearly 35 my, allowing ample opportunities for such changes to have occurred.

11.5 The Origin of Birds

Birds are usually regarded as being archosaurs and, therefore, a sister group of the crocodiles (Benton 2004). This view originated with E. Haekel who, in 1866, claimed the mammals as the sister group of the remaining amniotes. Haekel was supported by T.H. Huxley and Ray Lankester in 1870. A contrary opinion, however, has been expressed by Brian Gardiner (1982, 2002) who argued that crocodiles share their ancestry with both birds and mammals – having six synapomorphies with birds (including an eustachian tube that passes through the base of the cranium, a gizzard, and reduction of the fifth toes to the metatarsal) and two synapomorphies with mammals (calcaneal tubers and secondary palate). The fossil record of dinosaurs shows all the evolutionary changes that took place between crocodiles and birds (Gardiner 1993, 2002; Sect. 7.8).

11.5.1 Feathers

Among the more remarkable fossil discoveries of recent years have been those of small Upper Jurassic theropod dinosaurs which, although unable to fly, yet possessed feathers. One of these, *Sinosauropteryx* (Fig. 126a), has already been mentioned as being closely related to *Compsognathus* (Sect. 11.3.1). Another was *Protarchaeopteryx* (Fig. 126b), a typical coelurid but very similar to the earliest bird *Archaeopteryx* (Fig. 126c). *Protarchaeopteryx* had wrists that could twist to seize prey, elongated arms and hands with sharp claws. There was a clump of feathers at the end of the tail, which might perhaps have been used for display, and feathers on the forelimbs (Milner 2002). It seems highly

Fig. 126a–c. Dino-birds and *Archaeopteryx* the first bird (Upper Jurassic). **a** *Sinosauropteryx* (length ca. 1 m), **b** *Protarchaeopteryx* (length ca. 1 m), **c** *Archaeopteryx* (length ca. 37 cm)

probable, even if unprovable, that both feathers and hairs evolved from reptilian scales. The integumentary structure covering the bodies of *Sinosauropteryx* and the other 'dino-birds' were simpler than true feathers. They were composed of a central rachis or spine from which the barbs branched, but they lacked the aerodynamic quality of avian feathers. The longest ones covering the body were only about 3 cm in length and were probably more suitable for insulation than for display. The discovery of feathered dinosaurs provides support for the hypothesis not only that the theropods were direct ancestors to the birds but also that some, if not all of them, were tachymetabolic (Sect. 7.6). At the same time, the caudal feathers of *Protarchaeopteryx*, like those of *Caudipteryx*, might well have been used for display, possibly in conjunction with the longer feathers attached to the forearms.

11.5.2 Flight

The origin of flight in birds has aroused considerable speculation over the years. As Norman (1985) pointed out, there have been essentially two main schools of thought: flight evolved either in fast-running dinosaurs such as coelurosaurs or from gliding tree-dwellers. Leading advocates of the first of these hypotheses were Samuel Williston in 1879, Francis von Nopsca in 1907 and J.H. Ostrom in 1974. Williston suggested that gradual lengthening of the outer fingers and greater development of the scales might have aided a bipedal cursorial dinosaur in running. The "change of scales to feathers would have been easy. The wings must first have been used in running, next in leaping and descending from heights, and finally in soaring." Williston's hypothesis enjoyed little success among his contemporaries, and the following year O.C. Marsh proposed that the power of flight probably originated among small arboreal forms of reptilian birds. Nopsca's idea was similar to that of Williston, but more detailed and aerodynamically absurd, as Feduccia (1980) has clearly explained.

Reasoning from the anatomy of *Archaeopteryx* and from the features of dinosaurs contemporary with it, Ostrom (1974a) argued that the first feathers of tachymetabolic dinosaurs served as a thermoregulatory pelt. Enlargement of the primordial primaries and secondaries on the prey-catching forelimbs would have transformed these into "large, continuous trapping surfaces – or natural insect nets – activated by powerful ventral adductor muscles (the pectoralis group). These adaptations were admirably preadaptive for active, flapping flight." The desert islands in which *Archaeopteryx* lived did not contain any tall trees, and some scientists have suggested that perhaps it ran after flying insects, leapt to catch them in mid-air and then stayed aloft, flapping its wings. Its breast bone was too small to anchor strong muscles and it could not have flown at all well. During the Cretaceous period, however, many lines of birds with well-developed powers of flight evolved. Nevertheless, they died out some 5 my before the end of the Mesozoic Era.

Several important morphological, functional and physiological systems that are characteristic of modern birds were acquired only late in avian history. Chief among these was the modern pattern of hind limb kinematics – flexion of the knee and tibio-tarsal displacement, rather than the ancestral theropod pattern of hip extension. This modern pattern involves extensive retraction of the femur during each stride. It is probable, too, that throughout the evolution of the theropods, reduction in caudal musculature was the principal agent implicated in this mechanism. The tail was decoupled functionally from the primitive pattern of kinematics and, instead, coupled with the flight apparatus. At the same time, the centre of gravity moved forward and the modern avian stance developed (Chiappe 1995).

The arboreal origin of flight, first proposed by Marsh in 1926 is now generally favoured by palaeontologists (Feduccia 1980; Norman 1985; Milner 2002; Benton 2004). Kenneth Dial, for instance, has very recently produced evidence to support the view that the first feathered wings were used, not to keep jumping dinosaurs in the air for longer, but to increase the traction between their feet and the substrate when they were clambering up slopes steeper than about 45°. The idea was tested by filming partridges (*Perdix perdix*) as they scrambled up steep slopes and measuring the g-forces exerted by the birds' feet on the ground. Even newly hatched partridge chicks were able to use their stubby, undeveloped wings to climb a 59° incline (cited in The Independent, 17 January 2003).

11.6 Conclusions

The discovery that dromaeosaurs and dino-birds were covered with downy fibres (Ji et al. 1998, 2001) proved that feathers evolved first for insulation and were only secondly used in flight (Sect. 7.8). Birds evolved from cursorial, bipedal, carnivorous, coelurid dinosaurs with feathered forelimbs, hind legs and tails. These took to the air first by using their forelimbs to give added traction when running up sloping branches and, later, by leaping and gliding from one branch to another. By the Middle Jurassic (165 mya), dino-birds had evolved feathers and, by the early Cretaceous (130 mya), were similar in this respect to modern birds (class Aves) descended from them. *Archaeopteryx* was probably an awkward flier but in Cretaceous times more modern-looking birds such as the piscivorous *Ichthyornis* and the flightless diver *Hesperornis* evolved. Toothed birds such as these died out as the Mesozoic Era drew to its close and were replaced by modern types. Hypotheses regarding the origin of birds have been reviewed by Currie and Padian (1997) among others (see also Feduccia 1999; Blount and Crowley 2001).

12 The Cretaceous Extinction

12.1 Introduction

Extinction is the inevitable concomitant of evolution. As a result of natural selection in changing environments, all existing species must evolve into other chronologically distinct species, or else die out. None remains indefinitely without modification. Extinction is the consequence either of failure to adapt sufficiently rapidly, or else it occurs because the ecological niche of a species has disappeared. It can be caused by a combination of shortage of food, predation and disease, competition, and/or chance events (Hallam and Wignall 1997). There has long been a tendency to attribute the extinction of larger taxa, such as the dinosaurs, to specific catastrophic factors. As a result, incorrect or misleading conclusions have not infrequently been reached (Cloudsley-Thompson 2001).

When the Tertiary era dawned, none of the dinosaurs other than the birds remained (Charig 1989; Benton and Harper 1997; Benton 2004). The mammals were only just beginning to dominate the land and were therefore not serious competitors with them. Any explanation of dinosaur extinction must elucidate not only why many marine taxa, both invertebrate and vertebrate, also disappeared while the crocodilians and chelonians, as well as other reptiles, birds and mammals, survived (Benton 1989). This subject has attracted a tremendous amount of interest in recent years and no attempt will be made to review the voluminous literature. Instead, I shall merely try to indicate the views of many biologists – which have been generally swamped by those of some geologists and their followers among the general public.

12.1.1 Mass Extinctions

Many phenomena, which have recently been regarded as mass extinctions, are now recognised to be clusters of episodes of extinction only roughly associated in geological time. As pointed out by Raup (1992), an analysis of the extinction-intensity curve shows extinction intensities to be distributed continuously. At the same time, most authorities agree that no fewer than five genuine major extinctions are apparent in the fossil record. These occurred at the end of each of the Ordovician, Devonian, Permian, Triassic and Cretaceous periods (Table 3). The Cretaceous–Tertiary extinction during the Maastrichian stage at the close of the Mesozoic Era is known to geologists as K-T (*Kreta* is the Greek for chalk) to distinguish the word Cretaceous from Cambrian or Carboniferous. It was not the largest of the major extinctions although it has attracted by far the greatest popular interest. The reptilian groups or taxa that died out then included the mosasaurs, plesiosaurs, dinosaurs (except birds) – including the ankylosaurs, ceratopsians, coelurosaurs, hadrosaurs, hypselophodonts, tyrannosaurs and others (Archibald 1996) – but the ichthyosaurs, most pterosaurs, the pliosaurs and many other taxa had disappeared long before the K-T boundary.

The dinosaurs did not all become extinct at the same time but, from a peak of about 120 genera, numbers dwindled steadily during the final 300,000 years of the Mesozoic Era when the remainder finally vanished. Giant sauropods, including *Diplodocus* (Fig. 66), *Brachiosaurus* (Fig. 109) and *Apatosaurus* (Fig. 110), had become extinct long before the conclusion of the Cretaceous, while ceratopsians, such as *Triceratops* (Fig. 72) and the carnosaur *Tyrannosaurus* (Fig. 123) were among the last dinosaurs on earth (Weishampel et al. 1990; Currie and Padian 1997; Farlow and Brett-Surman 1997).

12.1.2
The Permian Extinction

The largest mass extinction of all time took place at the end of the Permian period 250 mya (Table 3). Of 45 families of animals present during the last 5 my of the Permian, no less than 36 – a loss of 75% – died out. These included 10 families of basal tetrapods (mostly anthracosaurs) and 17 families of therapsids including the gorgonopsians, the last dinocephalians and most of the dicynodonts. Only 12 families of tetrapods survived. Those that disappeared included a broad array of ecological types from large to small, herbivores and carnivores. At the same time, 50% of marine invertebrate families died out (Benton 2003, 2004).

Not surprisingly, the possible cause or causes of the mass extinction are numerous and have been much disputed. Asteroids and meteorites have been blamed, but there was no evidence of impact, and little shocked quartz (Sect. 12.4.1) was found before 2001. In that year, Luann Becker et al. reported that they had measured quantities of helium and argon trapped in large molecules of carbon (known as fullerenes) in geological samples from Hungary, China and Japan. These gases were identical to those previously derived from meteorites, and the authors argued that the Permo-Triassic boundary fullerenes must have come from impact with a meteorite – a claim that has been hotly disputed. Union of the elements of the supercontinent Pangaea (Sect. 2.2) had begun to occur, however, so there must have been considerable tectonic activity at the time. Flood basalt eruptions were undoubtedly taking place in Siberia in a region known as the Siberian Traps. During the course of 1 my, ca. 3 million km^3 of lava erupted over a large area, so huge amounts of smoke, carbon dioxide and sulphur dioxide would have been discharged into the atmosphere. The initial reduction in light intensity engendered a drop in temperature, which was followed as the darkness cleared by a rebound to ca. 5 °C above the original ambient temperature. This in itself would probably have been insufficient to explain the mass extinction, but it could have triggered a further rise of ca. 4–5 °C through the release of methane gas from the methane hydrate that accumulated on the continental margins (Hallam and Wignall 1997; Benton 2003). The same argument applies to the volcanic activity that preceded the Cretaceous extinction (Sect. 12.2). In each case, impact with a meteorite or bolide might perhaps have provided the *coup de grâce* for certain taxa.

12.1.3
The Triassic Extinction

A number of extinctions occurred at, or near to the close of the Triassic period (Table 3), and the end-Triassic event, based on a number of families that died out per million years, was almost as extensive as the K-T extinction. In fact, if the criterion used is the percentage of genera per million years, the extinction peak is even greater than that at the end of the Cretaceous. This, however, is made particularly apparent by the short time-span of the most recent Triassic unit. Estimates of its age range from 1–4 my, and are based on the extinction of marine invertebrate animals (Benton 1997a). As with other mass extinctions, the main possible causes involved are climate change, volcanism, bolide impact and changes in sea level. The latter would have been associated with anoxia. The evidence for each of these hypotheses has been reviewed by Hallam and Wignall (1997), but no conclusion was reached as to which is the more plausible.

12.2
Terminal Mesozoic Events

As a result of plate tectonics and continental drift during the last few million years of the Cretaceous period, mountains were forced upwards (the Laramide revolution), volcanoes erupted throughout the world (the Deccan plateau of India was formed by volcanic action), plateaux replaced floodplains and swamps, and the seas retreated. Many authorities believe that a period of global cooling occurred then, as a result of the reduction of solar heat by sulphurous gases; others that, due to increased carbon dioxide and methane in the atmosphere creating a 'greenhouse' effect, the temperature of the world increased.

During the Jurassic period, climatic conditions favoured a bloom of calcareous algae in the ocean. This must have depleted the atmosphere of carbon dioxide (CO_2) and the seas of calcium and bicarbonate ions. In deeper water, a zone of calcium carbonate ($CaCO_3$) under saturation probably developed. If this moved to the surface during the Late Cretaceous (Hallam and Wignall 1997), it would have destroyed the phytoplankton, triggered an ecological collapse, which is recorded as a gap of at least a million years in calcareous sedimentation, and engendered a consequent increase in atmospheric CO_2 resulting in global warming. Shortage of basic food materials might then have been responsible for increased competition and reduction in the numbers of molluscs, fishes and other marine animals upon which top predators, the mosasaurs and plesiosaurs, depended. Consequently, these too would have become extinct. The decline of the ammonites, inoceramid and rudist molluscs had, however, begun at least 6 my before the K-T boundary. In contrast, plant communities did not experience massive exterminations on land, although they did suffer some disturbances (as pointed out by Spicer and Parrish 1990 and by many other authors). The earlier extinction of the pterosaurs has already been discussed (Sect. 6.7.1).

Oxygen isotope studies of plankton and benthic Foraminifera suggest long-term cooling of the seas during the final Maastrichtian stage of the Late Cretaceous period. At the same time, there is evidence for K-T boundary warming in some parts of the world, just before the cooler period that may have followed. If temperatures did increase, a decrease in the solubility of oxygen might have reduced the amount of this gas in the oceans. There are deposits of black shale, which is caused by anoxic conditions, at the K-T boundary in several regions of the world (Hallam and Wignall 1997).

12.3 Dinosaur Extinction: Improbable Hypotheses

Numerous different hypotheses have been evoked to explain dinosaur extinction. Over 50 were listed by Glenn L. Jepson in 1964, and several more have been added since by M.J. Benton (1990a) and others (Fastovsky and Weishampel 1996). They include climatic deterioration; nutritional problems; disease; parasites; internecine fighting; anatomical or metabolic disorders such as slipped vertebral discs, malfunction or imbalance of hormone and endocrine systems etc.; racial senility; evolutionary drift into overspecialisation; changes in the composition and pressure of the atmosphere; widespread anoxia; too much or too little atmospheric carbon dioxide; poisonous gases including methane that destroyed the ozone layer and created an excessive rise in temperature; volcanic dust; excessive oxygen from plants; comets; meteorites; mammals preying on dinosaur eggs; extermination of herbivorous forms by carnivores, which then starved; temperature-induced changes in the sex ratios of embryos; fatal mutations caused by increased cosmic radiation – perhaps resulting from the explosion of a supernova; mountain building; fluctuation of gravitational constants; shifts of the poles; changes in the rotation of the earth; extraction of the moon from the Pacific basin; drainage of swamps and lake environments; the small size of dinosaur brains and the consequent stupidity of the animals; gigantism; suicidal psychoses and others even more far-fetched. Many of these are obviously untenable (Norman 1991), and only the most plausible suggestions will be considered here.

Although they blend into one another, hypotheses which might account for the K-T extinction tend to fall into two groups – those which attribute extinction to a sudden catastrophic event, and those which propose a gradual change over a much longer time. Extinction over a period of 100,000 years would appear to be very sudden in the palaeontological record! The idea of an extraterrestrial cause, such as fall out from a comet or impact with a bolide (a meteoric fireball) has a dramatic appeal to the popular imagination. The media are continually exploiting it, while Luis Alvarez (Alvarez et al. 1980, 1984) and James Lovelock, among many others, have been won over by it (Frankel 1999). Expertise in one subject does not, however, necessarily bestow insight into a completely different one. Palaeontologists such as David Archibald, Michael Benton, Alan Charig, Peter Dodson and Beverly Halstead, as well as most ecologists, tend to favour or have favoured more synthetic gradualist hypotheses based on careful consideration of all the evidence that can be accumulated from geology, meteorology, palaeontology and biology. These hypotheses may be less dramatic than those that evoke cosmic cataclysms alone but, because they are able to encompass several lines of evidence, they are more likely to approximate to the truth. As N. MacLeod has recently pointed out, Alvarez and his coworkers (1980, 1984) did at least move the question of extraterrestrial causes for mass extinction into the mainstream of science from which it had been excluded for some 150 years.

12.4 Cataclysmal Hypotheses

Geophysical surveys reveal that the earth's magnetic fields have become reversed on a number of occasions. Some physicists believe that this would have caused the van Allen belt to collapse and allowed cosmic rays and other radiation, normally deflected by the poles, to strike the earth. This can be ruled out as a cause of dinosaur extinction, however, because many marine forms were also profoundly affected despite the fact that a thin layer of water will act as a shield from cosmic rays. The radiation hypothesis cannot, therefore, be substantiated because many of the wrong creatures died.

Several authors have suggested that radiation from the explosion of a supernova, mostly in the form of X-rays, would be absorbed by the ozone layer at the ionosphere of the earth's outer atmosphere. The turbulence generated would have altered dramatically the property of the atmosphere to retain heat. While the drastic effects of a nearby supernova explosion are not in doubt, there is no evidence that one did occur at the end of the Cretaceous period and no explanation of why, if it did, its effects should have operated in the selective manner necessary to produce the observed results. The same objections apply to the suggestion of Harold Urey in 1973 that collision with a comet could have been responsible.

12.4.1 Impact with a Meteorite or Bolide

The discovery of a layer of iridium in samples of clay from the K-T boundary in Italy was interpreted by Alvarez et al. (1995) as having been caused by impact with a giant meteorite or bolide. Iridium is a rare element, often associated with extra-terrestrial bodies. It has subsequently been discovered at the K-T boundary in several other parts of the world. The environmental effects postulated by supporters of the bolide impact theory include severe reduction in light intensity due to a cloud of dust, meteorite debris and smoke from the wildfires engendered. In addition, temperatures would have dropped and acid rain fallen. In the original scenario, it was suggested that the earth had been plunged into darkness for 3 years. Toon et al. (1997), however, questioned whether dust from impact with an asteroid could have remained in the atmosphere for so long. Although dramatic sunsets after Krakatoa's eruption lasted for 3 years, Toon thought it probable that the dust would have fallen within 3 months at most, and that the striking sunsets observed after that time must have been due to sulphurous gases from the volcano. Alvarez (1997) therefore revised the period of darkness that he had postulated to one of between 3 and 6 months.

Even so, most biologists would find it difficult to accept the possibility that either planktonic organisms or, indeed, many terrestrial organisms could have survived for so long without light. It has been calculated that the microfauna of modern seas would consume all their food reserves within 10 to 100 days from the beginning of a blackout (Wilford 1985). According to some authorities, impact with a meteorite would have created gigantic tidal waves or tsunami, raised the temperature of the sea, and generated a super-hurricane or hypercane which would have blown dust and water high into the stratosphere, blacking out the sun for several months, reacting with atmospheric gases and wrecking the ozone layer.

The meteorite hypothesis in no way explains the precise mechanism of the extinction, nor its apparently random selectivity. Was it simply the effect of prolonged darkness or was there environmental cooling as a result of the shielding effect of dust in the atmosphere? Was there environmental warming through the greenhouse effect, and did the impact and resulting fireball produce torrents of acid rain? Perhaps several of these factors were combined (Norman 1991). Frankel (1999) attempted to explain the selectivity of the K-T extinction by postulating mass killing of the surface plankton and survival of benthic forms. Partial replacement of flowering plants by pteridophytes might, he suggested, have taken place on land accompanied by the survival of several invertebrate but only of small vertebrate animals.

Evidence for a meteorite impact is, however, boosted by the presence in the K-T layer of shocked quartz grains (in which the crystal structure is disrupted by fracture planes that occur at extremely high pressures as would have been caused by an impact; Archibald 1996). Finally, the presumed impact crater itself has been located off the coast of Yucatan, Mexico, near Chicxulub (Raup

1992; Frankel 1999). Some scientists (mostly with a geological rather than palaeontological background) unquestioningly assume that impact from a bolide must have been the cause of the final Cretaceous extinction. A statistical correlation does not, however, necessarily prove a causal connection (Raup 1989) and most authorities are far more cautious.

The precision of dating events that occurred many millions of years ago is, at best, in the order of many thousands of years. It is, therefore, impossible to determine from palaeontological evidence whether the K-T extinctions took place well before any asteroid impact, contemporaneously with it, or some time afterwards. Moreover, if a bolide were indeed responsible for the Cretaceous extinctions, it is likely that the time interval during which the various species died out would not have exceeded a year or two. It is not possible to be certain, because particular groups of organisms became extinct at the K-T boundary in one place, that they did so everywhere. Sudden disappearance could well be an illusion created by sampling error (McGowan 1991). There is also considerable evidence that the dinosaurs died out at different times in different places, sometimes fossils post-date the event that produced the iridium anomaly (Charig 1989). It is ironical that, of all the groups that became extinct, the Dinosauria is probably the one with the poorest fossil record!

A massive extraterrestrial disaster would certainly have made conditions on earth even more unfavourable to the dinosaurs and other large reptiles that were already declining at that time. Extensive volcanic activity might, however, have been responsible for much of the iridium anomaly (McGowan 1991): alternatively, a change in the sedimentation rate of seawater could have engendered an abnormal concentration of iridium. In any case, the Chicxulub bolide could well have been considerably smaller than of a 6-mile diameter as estimated by Alvarez and his colleagues in 1980 (see discussion in Frankel 1999).

12.5 Gradualist Hypotheses

The gradualist hypotheses favoured by many biologists and palaeontologists are even more numerous than the cataclysmic ones. Although many of those listed earlier (Sect. 12.2) are clearly nonsense, some unusual claims do merit consideration. For instance, Croft (1982) argued very convincingly that the dinosaur eye might well have lacked the protein β-crystalline which is present in all modern endothermal animals but not in reptiles. This protein is both photostable and thermally stable. Its absence would have rendered the dinosaurs particularly susceptible to the development of premature cataract blindness and this, over a long period, could have contributed to their extinction. Indeed, the body projections from the skulls of both ceratopsians and theropod carnosaurs might, in addition to their other functions, have served to shade the eyes from sunlight. Milne (1991) and others have, however, criticised this idea on the grounds that is unlikely that every dinosaur would have been selectively affected in this way. (The same objection also applies to hypotheses in which parasites, disease, mutations and so on are considered in isolation.) Furthermore, if the dinosaurs had evolved homeothermy, they might also have evolved β-crystalline.

12.5.1 Longevity, Injuries and Disease

Nothing, living or non-living, can persist indefinitely. In the days of the transatlantic liners, the average 'life' expectancy of a glass tumbler was about 1.5 crossings. Presumably, few tumblers would have survived for more than five or six crossings although potentially they could last for centuries. Whatever the potential life of a dinosaur, most individuals would have been killed by predators, injuries or disease long before they died of old age. Growth rings in certain bones and teeth suggest that some of the giant sauropods may have lived for at least 120 years. Perhaps some even reached 200 years, but there is no way of finding out (Lambert et al. 1992). During Mesozoic times, the earth was spinning faster on its axis than it does now, so the days were shorter and there were more of them in a year.

The greatest authentic age reached by a living reptile is over 150 years. This reptile was a giant tortoise (*Geochelone gigantea*) brought from the Seychelles to Mauritius in 1776 when fully mature. It was accidentally killed when it fell through a gun emplacement in 1918(!) and its actual age has since then been estimated at not less than 180 years. The greatest age recorded for a crocodilian is only 66 years.

Many Mesozoic reptile fossils show bones broken by falls or when fighting, or damaged by infections or arthritis. Dinosaur palaeopathology has been reviewed by Rothschild (1997). Evidence for injuries and disease is restricted mainly to abnormalities of bone growth, erosion of articular surfaces due to cartilage defects and fractures that have begun to heal. Breakages less than a few weeks old would probably not be recognisable as having occurred during life. Although osteoarthritis has traditionally been considered to be a common disorder of dinosaurs, signs of this, as well as of dental pathology, are extremely rare in dinosaur skeletons. Gout has, however, been identified in the finger and toe bones of *Tyrannosaurus*, fusion of the cervical vertebrae in ceratopians, and tail tip fusion in hadrosaurs. Examples of tumours are very problematic. Eggs of *Hypselosaurus* with multilayered shells have also been found. Stress is known to cause thinning in the shells of birds' eggs by disturbing the hormones of the birds that are laying them. Erben et al. (1979) found progressive thinning in some Late Cretaceous dinosaur eggs, which they thought might indicate stress, but similar observations have not been made in the case of other species.

The factors that limit life span, such as increased susceptibility to infectious disease, do not usually affect the skeleton. Consequently, palaeopathology does not provide any clues as to the possible causes of extinction. Nevertheless, it is clear from the paucity of pathological evidence that the dinosaurs and the other Mesozoic reptiles must have been relatively healthy animals.

12.5.2 Nutritional Factors

By the end of the Mesozoic Era, an essentially modern type of flora had evolved on land. Gymnosperms, which had been the dominant plants, were gradually superseded by angiosperms, especially deciduous trees and shrubs, adapted to seasonal climates. Shortly before World War II, Cambridge biochemist Ernest Baldwin suggested that an enforced change in diet from tree ferns to deciduous plants might have caused constipation and other nutritional disorders for herbivorous dinosaurs. In 1976, Tony Swain argued that the ability of flowering plants to produce hydrolysable tannins, which are more efficient as feeding deterrents than the compounds present in lower plants, would have greatly reduced the amount of palatable food available to all herbivores. Furthermore, angiosperms synthesise many toxic alkaloids that are absent from most lower plants and are far less easily detected by extant reptiles than they are by mammals. Although these alkaloids might have been responsible for direct poisoning of some archosaurs, Swain concluded it to be more probable in the longer term that they would have interfered with several physiological activities and caused the observed thinning of ceratopsian eggshells (Sect. 12.5.1).

However, if toxic alkaloids appeared 120 mya with the first flowering plants, one wonders why the dinosaurs lingered on for a further 50 my before being struck down. Moreover, the incompatible suggestion has been made by Howard Hinton (1973) that the bright colours of many flowers – especially at the red end of the spectrum, invisible to insect pollinators – may have evolved as aposematic (warning) coloration. Mammalian herbivores lack colour vision while, as far as is known, all reptiles possess it. It is therefore possible that one of the chief selective pressures for the evolution of coloured flowers may have come from herbivorous dinosaurs. This could scarcely have occurred if they had been unable to detect the poisonous chemicals of which the flowers gave warning (Cloudsley-Thompson 1978).

12.5.3 Thermal Effects

For more than 100 my before their demise, the dinosaurs enjoyed equable climatic conditions. At the end of the Mesozoic Era, however, the seasons became accentuated, as indicated by the rise of the angiosperms with their deciduous foliage and overwintering seeds. In 1965, the Canadian palaeontologist L.S. Russell proposed that winter temperatures might eventually have fallen low enough to kill such land animals as could not conserve their body warmth and did not hibernate. This might explain why the dinosaurs died out, even though they may have been tachymetabolic (with a fast rate of chemical change; Sect. 7.6), as argued by Bakker (1987) since the 1960s, while reptiles that were not endothermal were able to survive. It seems more probable, however, that the summer heat would have been more unfavourable, as argued below – a hypothesis that fits with the recent discovery of dinosaur fossils near the poles. Russell's suggestion also fails to explain why the dinosaurs did not persist in equatorial regions, as the crocodilians did, nor why the theropods, many of which are known to have been feathered (Sect. 11.5.1), also died out.

Because chemical reactions are temperature dependent (Q_{10} effect), there is an evolutionary tendency towards systems that enable animals to maintain their body temperature at a constant level. It is now believed that the sail of the pelycosaur *Dimetrodon* (Fig. 10) may have been a device for absorbing solar heat in the cool Permian mornings, and for radiating excess heat during the hot part of the day (Sect. 7.5.2). In a similar way, according to Farlow et al. (1976), the dorsal plates of stegosaurs (Fig. 71) probably served as forced convection fins and had an important thermoregulatory function, as well as being concerned with intra-specific combat and anti-predator display (Sects. 7.5.2, 9.3.1).

The temperature at which the eggs of modern crocodilians and other reptiles are incubated affects the gender of the hatchlings. The possible influences of thermal sex determination in the K-T extinction has been discussed by Milne (1991), but it is almost unbelievable that the dinosaurs, with the gradual onset of seasons, did not evolve seasonal breeding cycles like other reptiles (Cloudsley-Thompson 1978).

Whether or not the dinosaurs were tachymetabolic (Sect. 7.6) is, I believe, largely immaterial as regards their extinction. The larger forms must have been homeothermic whether they were endothermal or not. Even the smaller species were too large to burrow and many may have been on the way to becoming birds. In contrast, modern reptiles are able to avoid inclement seasons by aestivating (summer dormancy) or hibernating in sheltered retreats. In addition, they utilise physiological thermoregulatory processes, such as emergency cooling of the body of tortoises through salivation and urination. The elongated shape and comparatively small size of most existing reptiles render prolonged homeothermy uneconomical, and even large pythons 'shiver' only when brooding their eggs (Cloudsley-Thompson 1999). The small surface-to-volume ratio of large animals, including most of the very last dinosaurs, would have rendered them susceptible to heat rather than to cold, a point overlooked by many non-biologists such as Milne (1991). Moreover, fermentation in the gut of the large herbivores would have generated a considerable amount of heat.

The thermal physiology of dinosaurs has already been discussed (Sect. 7.5). Gigantism is a useful strategy for reptiles in a stable, relatively warm climate, but thermal stresses caused by decreasing equability at the close of the Mesozoic might well have had harmful effects on large animals. Hair would have been as disadvantageous to a large dinosaur in summer as it would be to a modern African elephant or rhinoceros. Furthermore, the work of Clemens (1986) and others in recent years has shown that the last of the dinosaurs and other extinct reptiles resided not in the tropics but in temperate regions nearer to the poles.

Why did compensation for size not prevent excess metabolic heat from being generated whether the dinosaurs were endothermic or exothermic, especially if they were as active as we now believe them to have been? It is perhaps surprising that the more advanced vertebrates of today have evolved tachymetabolic homeothermy rather than enzyme systems that could have functioned equally well at lower temperatures. The list of relevant problems seems endless. Nevertheless, of the many factors that undoubtedly contributed to the ultimate lack of competitiveness and demise of the dinosaurs, thermal stress may well have been one of the more significant – especially in the summer (Cloudsley-Thompson 1978).

12.6 Conclusions

Few palaeontologists appear to be greatly concerned as to the causes of the extinction of dinosaurs and other Mesozoic taxa. Extinction is inevitable and it is more interesting to study how the archosaurs lived than why they died. 'The Age of Reptiles' began about 250 mya and ended some 185 my later. It seems probable from a biological viewpoint that the forms which became extinct then went, 'not with a bang but a whimper', to quote T.S. Eliot's *The Hollow Men* (published in 1925).

Many people now accept that there may have been a bolide impact about the time of the K-T boundary. There was also a considerable amount of volcanic activity, mountain building, marine transgression, recession and so on, associated with continental drift. The evidence supporting both catastrophic and gradualist explanations of the end-Cretaceous reptilian extinctions has been enumerated by Russell (1971), McGowan (1991), Milne (1991), Benton (1996), Fastovsky and Weishampel (1996), Russell and Dodson (1997) and many others.

Among these, Archibald (1996, 1997) believed marine regression, extraterrestrial impact and, to a lesser extent, massive volcanism to have been responsible. To biologists and ecologists, who realise that natural selection operates at several different levels and on many morphological and physiological functions simultaneously, it seems impossible that the numerous selective proximal extinctions that took place could have been the result of a single cause, however dramatic, or even of a limited number of causes. Benton (2004) summed up the matter as follows: "available killing models are either biologically unlikely, or are too catastrophic ... 70–75% of families survived the K-T event, many of them seemingly entirely unaffected! Some palaeontological evidence supports the view of instantaneous extinction, but the majority indicates longer-term extinction over 1–2 my." The event can be understood better when the time scale is broadened (Fig. 127).

Multiple proximal factors ultimately associated with tectonic activity, and operating differently on land and in the oceans, must surely have been mainly responsible for the K-T extinction. Impact with a bolide might also have contributed to the extinctions. For a wide variety of reasons, changing environmental

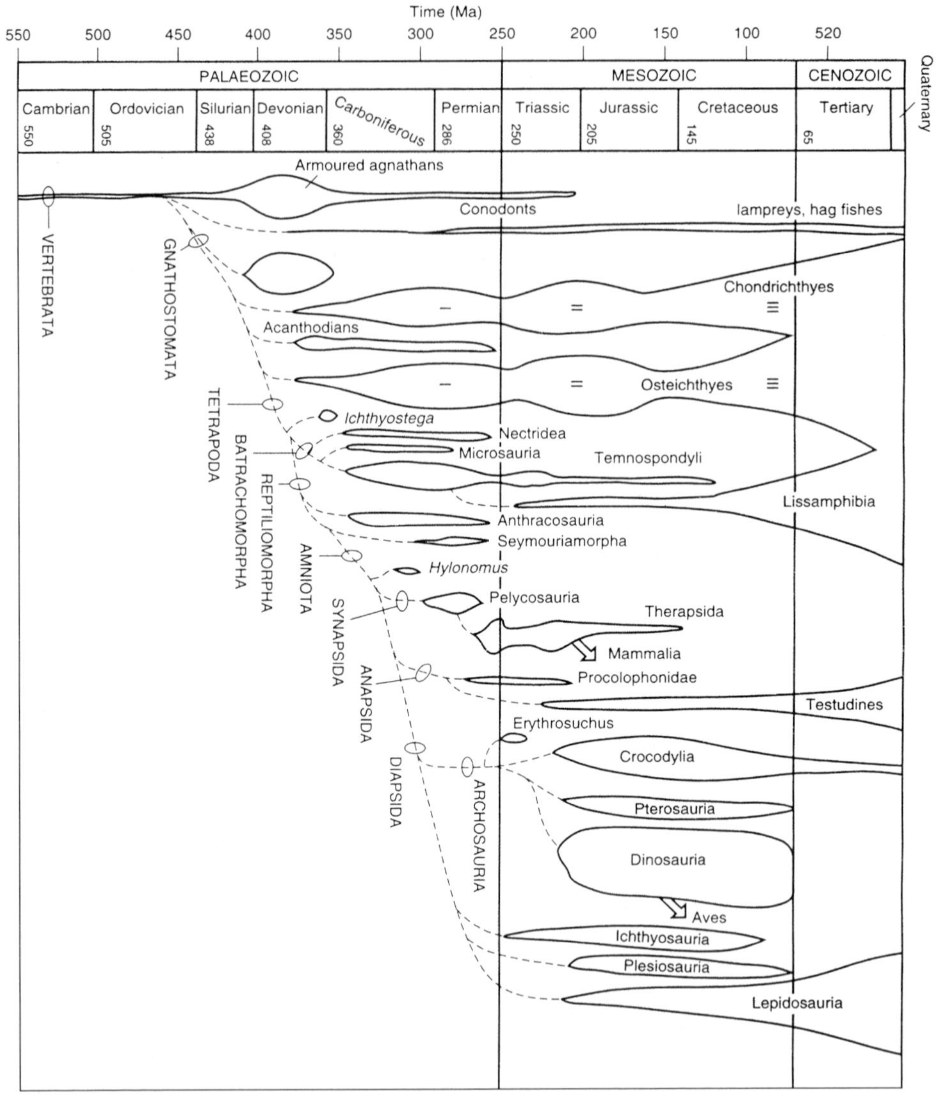

Fig. 127. Phylogeny of the major groups of vertebrates. (Benton and Harper 1997)

and climatic conditions tend to make certain groups somewhat less competitive than others and, consequently, they are replaced. It will probably be more profitable in the future to study the recovery of the taxa that outlived the Mesozoic Era than to investigate the disappearance of those that did not!

E Epilogue

Fossils of over 800 genera of dinosaurs have been found and described – usually on the basis of only a few teeth or bones. More are identified every year. Most of these genera are based on a single species, which suggests that thousands more genera and species are waiting to be discovered. Numerous others may never have been fossilised. Yet the dinosaurs were only one of the major taxa of Mesozoic reptiles. This gives some indication of the plethora of reptile species that must have evolved and become extinct during the Era. Each of these must have been adapted to a specific habitat at a particular moment during a time span of about 185 my. The smallest known dinosaur was only 60 cm in length and weighed about 3·5 kg, the largest was over 30 m long and weighed some 60 tonnes, but all that these scaly-skinned reptiles left behind were fossilised skeletons and trackways. The feathered dinosaurs are with us still in the form of birds.

Our solar system began some 4,550 mya when a star collapsed and exploded, but life did not appear for another 1,000 my. Even so, the period during which vertebrates have existed represents only a tiny fraction of the time during which the world has supported life, and it was not until around 540 mya that the chordates arose. The first reptiles evolved during the Middle Cambrian and the pelycosaurs diversified in the Early Permian. The Mesozoic Era (250–65 mya) lasted for 185 my. Some idea of the length of time involved can be gained from the thickness of the Upper Cretaceous chalk deposits of southern England and Europe. These were formed from the shells of microscopic marine organisms over many millions of years. Crude oil, too, was formed from organic remains by the action of heat and pressure over comparable millions of years. Again, when Pangaea broke up and continental drift began, at a rate of a centimetre or so per year, organic matter in sedimentary rocks was forced downward by movement of the tectonic plates, eventually to be melted and finally expelled in the form of carbon dioxide in the gaseous emissions of modern volcanoes.

Many of the Mesozoic reptiles were more spectacular and exciting than any animals that had appeared before that time or have existed later. Different forms evolved in response to particular environmental conditions, and died out when these conditions changed. When similar conditions returned, they were replaced by ecological equivalents, usually not closely related to them. Such considerations enhance the fascination afforded by the study of ancient life in general and of Mesozoic reptiles in particular.

B Bibliography

The literature on Mesozoic reptiles is vast. The present list of publications includes by no means all those upon which the text has been based. For the sake of conciseness, reviews and recent books, particularly when well illustrated, have been cited in preference to original research publications. The names of artists are given when listed on title pages. Especially significant as sources of reference are the following: Thomas and Olson (1980); Kemp (1982); Hotton et al. (1986); McGowan (1991); Wellnhofer (1991); Archibald (1996); Dodson (1996); Callaway and Nicholls (1997); Currie and Padian (1997); Farlow and Brett-Surman (1997); Hallam and Wignall (1997); Ellis (2003); Benton (2004).

Abel O (1925) Geschichte und Methode der Reconstruktion vorzeitlicher Wirbeltiere. Fischer, Jena, 327 pp

Alderton D (1991) Crocodiles and alligators of the world. Blandford, London, 190 pp

Alexander RMcN (1971) Size and shape. Studies in biology 29. Arnold, London, 59 pp

Alexander RMcN (1985) Mechanics of posture and gait of some large dinosaurs. Zool J Linn Soc 83:1–25

Alexander RMcN (1989) Dynamics of dinosaurs and other extinct giants. Columbia Press, New York, vii + 167 pp

Alexander RMcN (1997a) Engineering a dinosaur. In: Farlow JO, Brett-Surman MK (eds) The complete dinosaur. Indiana Univ Press, Bloomington, pp 414–425

Alexander RMcN (1997b) Size and scaling. In: Currie PJ, Padian K (eds) Encyclopedia of dinosaurs. Academic Press, San Diego, pp 665–667

Alvarez LW, Alvarez W, Asaro F, Michel HV (1980) Extra-terrestrial cause of the Cretaceous-Tertiary extinction. Science 208:1095–1108

Alvarez W (1997) *T. rex* and the crater of doom. Princeton Univ Press, Princeton, 185 pp

Alvarez W, Alvarez L W, Asaro F, Michel H V (1984) The end of the Cretaceous: sharp boundary or gradual transition? Science 223:1183–1186

Alvarez W, Claeys P, Kieffer SW (1995) Emplacement of Cretaceous-Tertiary boundary shocked quartz from Chicxulub Crater. Science 269:930–935

Archibald J D (1996) Dinosaur extinction and the end of an era. What the fossils say. Columbia Univ Press, New York, xviii + 237 pp

Archibald JD (1997) Extinction, Cretaceous. In: Currie PJ, Padian K (eds) Encyclopedia of dinosaurs. Academic Press, San Diego, pp 221–230

Attridge J, Crompton AW, Jenkins FA (1985) The southern African Liassic prosauropod *Massospondylus* discovered in North America. J Vertebr Paleontol 5:128–132

Augusta J (1956) Prehistoric animals (Illstr Burian Z) (Transl Hort G). Spring Books, London, 47 pp, 60 pls

Augusta J (1961) Prehistoric reptiles and birds (Illustr Burian Z) (Transl Schiel M). Hamlyn, London, 103 pp

Axelrod DI, Bailey HP (1968) Cretaceous dinosaur extinction. Evolution 22:595–611

Bakker RT (1968) The superiority of dinosaurs. Discovery (New Haven) 3(2):11–22

Bakker RT (1971a) Dinosaur physiology and the origin of mammals. Evolution 25:836–858
Bakker RT (1971b) Ecology of the brontosaurs. Nature 229:172–174
Bakker RT (1972) Anatomical and ecological evidence of endothermy in dinosaurs. Nature 238:81–85
Bakker RT (1974) Dinosaur bioenergetics – a reply to Bennett and Dalzell, and Feduccia. Evolution 28:497–503
Bakker RT (1975) Dinosaur renaissance. Sci Am 232(4):58–78
Bakker RT (1978) Dinosaur feeding behaviour and the origin of flowering plants. Nature 274:661–663
Bakker R (1987) The dinosaur heresies. A revolutionary view of dinosaurs. Longman Scientific & Technical, Harlow Essex, 482 pp
Barsbold R, Osmólska H (1990) Ornithomimosauria. In: Weishampel DB, Dobson P, Osmólska H (eds) The Dinosauria. Univ California Press, Berkeley, pp 225–244
Becker L, Hunt AG, Bunch TE, Rampino M (2001) Impact event at the Permian-Triassic boundary; evidence from extra-terrestrial noble gases in fullerenes. Science 291:1530–1533
Bellairs A d'A, Attridge J (1975) Reptiles. Hutchinson, London, 240 pp
Bellairs A d'A, Cox CB (eds) (1976) Morphology and biology of reptiles. Linnean Society symposium series no 3. Academic Press, London, xi + 290 pp
Benedict FG (1932) The physiology of large reptiles. Carnegie Publ, no 425, Washington, 539 pp
Bennett AF (1974) A final word. Evolution 28:503
Bennett AF, Dawson WR (1976) Metabolism. In: Gans C, Dawson WR (eds) Biology of the Reptilia, 5 Physiology A. Academic Press, London, pp 127–223
Bennett AF, Dalzell B (1973) Dinosaur physiology: a critique. Evolution 27:170–174
Bennett AF, Ruben JA (1986) The metabolic and thermoregulatory states of therapsids. In: MacLean PD, Roth JJ, Roth EC (eds) The ecology and biology of mammal-like reptiles. Smithsonian Inst, Washington, DC, pp 207–218
Bennett SC (1987) Sexual dimorphism in the pterosaur *Pteranodon*. J Vertebr Paleontol 7 [Suppl to no 3]:9
Bennett SC (1990) A pterodactyloid pterosaur from the Santana Formation of Brazil: implications for terrestrial locomotion. J Vertebr Paleontol 10:80–85
Bennett SC (2003) New crested specimens of the Late Cretaceous pterosaur *Nyctosaurus*. Palaeontol Z 77:61–75
Benton MJ (1983) Dinosaur success in the Triassic: a noncompetitive ecological model. Q Rev Biol 58:29–55
Benton MJ (1985) Classification and phylogeny of the diapsid reptiles. Zool J Linn Soc 84:97–164
Benton M J (1989) Mass extinctions among tetrapods and the quality of the fossil record. Philos Trans R Soc Lond B 325:369–386
Benton MJ (1990a) Scientific methodologies in collision. The history of the study of the extinction of the dinosaurs. Evol Biol 24:371–400
Benton MJ (1990b) Origin and interrelationships of the dinosaurs. In: Weishampel DB, Dobson P, Osmólska H (eds) The Dinosauria. Univ California Press, Berkeley, pp 11–29
Benton MJ (1996) The Penguin historical atlas of the dinosaurs. Penguin Books, London, 144 pp
Benton MJ (1997a) Extinction, Triassic. In: Currie PJ, Padian K (eds) Encyclopedia of dinosaurs. Academic Press, San Diego, pp 230–236
Benton MJ (1997b) Origin and early evolution of dinosaurs. In: Farlow JO, Brett-Surman MK (eds) The complete dinosaur. Indiana Univ Press, Bloomington, pp 204–215
Benton MJ (2003) When life nearly died. The greatest mass extinction of all time. Thames and Hudson, London, 336 pp
Benton MJ (2004) Vertebrate palaeontology, 3rd edn. Blackwell, Oxford, xi + 455 pp
Benton MJ, Harper DAT (1997) Basic palaeontology. Addison-Wesley, New York, 357 pp
Bird RT (1944) Did brontosaurs ever walk on land? Nat Hist 53:60–67
Blount M, Crowley M (eds) (2001) D K Encyclopedia of dinosaurs and prehistoric life (in association with the American Museum of Natural History). Dorling Kindersley, London, 376 pp
Bouvier M (1997) Dinosaur Haversian bone and endothermy. Evolution 31:449–450

Bramwell CD, Fellgett PB (1973) Thermal regulation in sail-lizards. Nature 242:203–205

Bramwell CD, Whitfield GR (1974) Biomechanics of *Pteranodon*. Philos Trans R Soc Lond B 267:503–581

Brett-Surman MK (ed) (2000) Dinosaurs. The ultimate guide to prehistoric life. Harper Collins, London, 265 pp

Brochu CA, Long J, McHenry C, Scanlon JD, Willis P (2000) A guide to dinosaurs. In: Brett-Surman MK (ed) Dinosaurs. The ultimate guide to prehistoric life. Harper Collins, London, pp 112–205

Brown DS (1981) The English Upper Jurassic Plesiosauroidea (Reptilia), and a review of the phylogeny and classification of the Plesiosauria. Bull Brit Mus (Nat Hist) Geol 35:253–347

Buchholtz E (1997) Paleoneurology. In: Currie PJ, Padian K (eds) Encyclopedia of dinosaurs. Academic Press, San Diego, pp 522–524

Buffetaut E (1989) Evolution. In: Ross CA, Garnett S (eds) Crocodiles and alligators. Welden Owen, McMahons Point NSW, Merehurst Press, London, pp 26–35

Callaway JM, Nicholls EL (eds) (1997) Ancient marine reptiles. Academic Press, San Diego, xlvi + 501 pp

Carroll RL (1997) Mesozoic marine reptiles as models of long-term, large-scale evolutionary phenomena. In: Callaway JM, Nicholls EL (eds) Ancient marine reptiles. Academic Press, San Diego, pp 467–489

Charig A (1976) Dinosaur monophyly and a new class of vertebrates. In: Bellairs A d'A, Cox CB (eds) Morphology and biology of reptiles. Academic Press, London, pp 65–104

Charig A (1979) A new look at the dinosaurs (Illustr Snowball P, Burrows R, Burrows C). Heinemann, London Melbourne Toronto. In association with the British Museum (Natural History), 160 pp

Charig AJ (1984) Competition between therapsids and archosaurs during the Triassic period: a review and synthesis of current theories. In: Ferguson MJW (ed) The structure, development and evolution of reptiles. Symp Zool Soc Lond 52:597–628

Charig AJ (1989) The Cretaceous–Tertiary boundary and the last of the dinosaurs. Philos Trans R Soc Lond B 325:387–400

Charig AJ, Attridge J, Crompton AW (1965) On the origins of the sauropods and the classification of the Saurischia. Proc Linn Soc Lond 176:197–221

Chiappe LM (1995) The first 85 million years of avian evolution. Nature 378:349–355

Chiappe L, Dingus L (2001) Walking on eggs. Discovering the astonishing secrets of the world of dinosaurs (Illustr Frankfurt N). Little Brown, London, 219 pp

Chin K (1997) What did dinosaurs eat? Coprolites and other direct evidence of dinosaur diets. In: Farlow JO, Brett-Surman MK (eds) The complete dinosaur. Indiana Univ Press, Bloomington, pp 371–382

Clark JM, Jacobs LL, Downs WR (1989) Mammal-like dentition in a Mesozoic crocodylian. Science 244:1064–1066

Clemens WA (1986) Evolution of the terrestrial vertebrate fauna during the Cretaceous-Tertiary transition. In: Elliott DK (ed) The dynamics of extinction. Wiley Interscience, New York, pp 63–86

Cloudsley-Thompson JL (1977) The size of animals. Patterns of progress 3. Meadowfield Press, Shildon Co Durham, 58 pp

Cloudsley-Thompson JL (1978) Why the dinosaurs became extinct. Patterns of progress 15. Meadowfield Press, Shildon Co Durham, 85 pp

Cloudsley-Thompson JL (1994) Predation and defence amongst reptiles. R & A, Taunton Somerset, viii + 138 pp

Cloudsley-Thompson JL (1999) The diversity of amphibians and reptiles. An introduction. Springer, Berlin Heidelberg New York, 234 pp

Cloudsley-Thompson JL (2001) Multiple factors in the reptile extinctions of the Cretaceous period. Biologist 48(4):177–181

Colbert EH (1962a) The weights of dinosaurs. Am Mus Nov 2076:1–16

Colbert EH (1962b) Dinosaurs. Their discovery and their world. Hutchinson, London, 288 pp

Colbert EH (1965) The age of reptiles (The World Naturalist). Weidenfeld and Nicolson, London, xiv + 228 pp
Colbert EH (1966) Evolution of the vertebrates. A history of the backboned animals through time, 3rd edn. Wiley, New York, xvi + 510 pp
Colbert EH (1968) Men and dinosaurs. Dutton, New York, 283 pp
Colbert EH (1993) Feeding strategies and metabolism in elephants and sauropod dinosaurs. Am J Sci 293A:1–19
Colbert EH, Cowles RB, Bogert CM (1946) Temperature tolerances in the American alligator and their bearing on the habits, evolution and extinction of the dinosaurs. Bull Am Mus Nat Hist 86:329–373
Collin R, Janis CM (1997) Morphological constraints on tetrapod feeding mechanisms: why were there no suspension-feeding marine reptiles? In: Callaway JM, Nicholls EL (eds) Ancient marine reptiles. Academic Press, San Diego, pp 452–466
Coombs WR Jr (1990) Behavior patterns of dinosaurs. In: Weishampel DB, Dobson P, Osmólska H (eds) The Dinosauria. Univ California Press, Berkeley, pp 32–62
Cott HB (1961) Scientific results of an inquiry into the ecology and economic status of the Nile crocodile (*Crocodilus niloticus)* in Uganda and northern Rhodesia. Trans Zool Soc Lond 29(4):211–356
Cowles RB (1940) Additional implications of reptilian sensitivity to high temperatures. Am Nat 74:542–561
Cowles RB (1958) Possible origin of dermal temperature regulation. Evolution 12:347–357
Croft LR (1982) The last dinosaurs. A new look at the extinction of the dinosaurs. Elmwood, Chorley Lancashire, 80 pp
Currie PJ (1997a) Theropoda. In: Currie PJ, Padian K (eds) Encyclopedia of dinosaurs. Academic Press, San Diego, pp 731–737
Currie PJ (1997b) Theropods. In: Farlow JO, Brett-Surman MK (eds) The complete dinosaur. Indiana Univ Press, Bloomington, pp 216–233
Currie PJ, Padian K (eds) (1997) Encyclopedia of dinosaurs. Academic Press, San Diego, xxx + 869 pp
de Ricqlès A (1974) Evolution of endothermy: historical evidence. Evol Theory 1:51–80
de Ricqlès A (1975) Les premiers vertébrés volants. Recherche 6:608–617
de Ricqlès A (1976) On bone histology of fossil and living reptiles, with comments on its functional and evolutionary significance. In: Bellairs A d'A, Cox CB (eds) Morphology and biology of reptiles. Academic Press, London, pp 123–150
de Ricqlès A (1980) Tissue structure of dinosaur bone. Functional significance and possible relation to dinosaur physiology. In: Thomas RDK, Olsen EC (eds) A cold look at the warm-blooded dinosaurs. Westview Press, Boulder CO, pp 103–239
Desmond AJ (1975) The hot-blooded dinosaurs. A revolution in palaeontology. Blond and Briggs, London, 238 pp
Dodson P (1974) Dinosaurs as dinosaurs. Evolution 28:494–497
Dodson P (1976) Quantitative aspects of relative growth and sexual dimorphism in *Protoceratops*. J Paleontol 50:929–940
Dodson P (1990) Sauropod paleoecology. In: Weishampel DB, Dobson P, Osmólska H (eds) The Dinosauria. Univ California Press, Berkeley, xxii + 733 pp
Dodson P (1996) The horned dinosaurs. A natural history (Illustr Barlow JD). Princeton Univ Press, Princeton, xiv + 346 pp
Ellis R (2003) Sea dragons: predators of the prehistoric oceans. Univ Kansas Press, Lawrence, xii +313 pp
Erben HK, Hoefs J, Wedepohl KH (1979) Paleobiological and isotopic studies of egg shells from a declining dinosaur species. Paleobiology 5:380–414
Evans SE (1982) The gliding reptiles of the Upper Permian (Drawing by Green D). Zool J Linn Soc 76:97–123
Farlow JO (1976) A consideration of the trophic dynamics of a late Cretaceous large-dinosaur community (Oldman Formation). Ecology 57:841–857

Farlow JO (1987) Speculations about the diet and digestive physiology of herbivorous dinosaurs. Paleobiology 13:60–72

Farlow JO (1993) On the rareness of big fierce animals: speculations about the body sizes, population densities and geographic ranges of predatory mammals and large carnivorous dinosaurs. Am J Sci 293A:167–199

Farlow J O (1997) Dinosaur energetics and thermal biology. In: Weishampel DB, Dobson P, Osmólska H (eds) The Dinosauria. Univ California Press, Berkeley, pp 43–55

Farlow JO, Brett-Surman MK (eds) (1997) The complete dinosaur (Art ed Walters R F). Indiana Univ Press, Bloomington, xiii + 753 pp

Farlow JO, Chapman RE (1997) The scientific study of dinosaur footprints. In: Farlow JO, Brett-Surman MK (eds) The complete dinosaur. Indiana Univ Press, Bloomington, pp 519–553

Farlow JO, Dodson P (1975) The behavioural significance of frill and horn morphology in ceratopsian dinosaurs. Evolution 29: 353–361

Farlow JO, Thompson CV, Rosner DE (1976) Plates of the dinosaur *Stegosaurus*: forced convection heat loss fins? Science 192:1123–1125

Fastovsky DE, Weishampel DB (1996) The evolution and extinction of the dinosaurs (Illustr Regal B). Cambridge Univ Press, Cambridge, xiv + 460 pp

Feduccia A (1973) Dinosaurs as reptiles. Evolution 27:116–119

Feduccia A (1974) Endothermy, dinosaurs and *Archaeopteryx*. Evolution 28:503–504

Feduccia A (1980) The age of birds. Harvard Univ Press, Cambridge, Mass, London, xi + 196 pp

Feduccia A (1999) The origin and evolution of birds, 2nd edn. Yale Univ Press, New Haven/Wiley, London, 478 pp

Fleming TH, Lips KR (1991) Angiosperms endozoochory: were pterosaurs Cretaceous seed dispersers? Am Nat 138:1058–1065

Frankel C (1999) The end of the dinosaurs. Chicxulub crater and mass extinctions. Cambridge Univ Press, Cambridge, xii + 223 pp

Galton PM (1970a) The posture of hadrosaurian dinosaurs. J Paleontol 44:464–473

Galton PM (1970b) Pachycephalosaurids – dinosaurian battering rams. Discovery (New Haven) 6:23–32

Galton PM (1985) Diet of prosauropod dinosaurs from the Late Triassic and Early Jurassic. Lethaia 18:105–123

Galton PM (1986) Herbivorous adaptations of Late Triassic and Early Jurassic dinosaurs. In: Padian K (ed) The beginning of the age of dinosaurs. Faunal change across the Triassic–Jurassic boundary. Cambridge Univ Press, Cambridge, pp 203–221

Gardiner BG (1982) Tetrapod classification. Zool J Linn Soc 74:207–232

Gardiner BG (1993) Haematothermia: warm-blooded amniotes. Cladistics 9:369–395

Gardiner BG (2002) Crocodile relationships. Linnean 18(2):33–40

Gillette D (1994) *Seismosaurus*. The earth shaker (Illustr Hallett M). Columbia Univ Press, New York, x + 206 pp

Godfrey SJ (1984) Plesiosaur subaqueous locomotion: a reappraisal. N Jahrb Geol Paläontol Monatsh 11:661–672

Goodwin MB, Johnson RE (1995) A new skull of the pachycephalosaur *Stygimoloch* casts doubt on head butting behavior. J Vertebr Paleontol 15(3, Suppl):32A

Gray J (1968) Animal locomotion. Weidenfeld and Nicolson, London, xi + 479 pp

Haack SC (1986) A thermal model of the sailback pelycosaur. Paleobiology 12:459–473

Hallam A, Wignall PB (1997) Mass extinctions and their aftermath. Oxford Univ Press, Oxford, 320 pp

Halstead LB (1989) Plesiosaur locomotion. J Geol Soc Lond 146:37–40

Halstead LB, Halstead J (1981) Dinosaurs. Blandford Press, Poole Dorset, 170 pp

Heath J E (1968) The origin of thermoregulation. In: Drake ET (ed) Evolution and environment. Yale Univ Press, New Haven, pp 259–278

Hinton HE (1973) Natural deception. In: Gregory RL, Gombrich HE (eds) Illusion in nature and art. Duckworth, London, pp 95–159

Hohnke LA (1973) Haemodynamics of the Sauropoda. Nature 244:309–310

Holtz TR Jr, Brett-Surman M K(1997) Taxonomy and systematics of the dinosaurs. In: Farlow JO, Brett-Surman MK (eds) The complete dinosaur. Indiana Univ Press, Bloomington, pp 92–106
Hopson JA (1977) Relative brain size and behaviour in archosaurian reptiles. Annu Rev Ecol Syst 8:429–448
Horner JR (1997) Behavior. In: Currie PJ, Padian K (eds) Encyclopedia of dinosaurs. Academic Press, San Diego, pp 45–50
Hotton N III (1980) An alternative to dinosaur endothermy. The happy wanderers. In: Thomas RDK, Olsen EC (eds) A cold look at warm-blooded dinosaurs. Westview Press, Boulderco, pp 311–350
Hotton N III, Maclean PD, Roth JJ, Roth EC (eds) (1986) The ecology and biology of mammal-like reptiles. Smithsonian Institute Press, Washington, DC, x + 326 pp
Hua S, Buffetaut E (1997) Part V: Crocodylia. Introduction. In: Callaway JM, Nicholls EL (eds) Ancient marine reptiles. Academic Press, San Diego, pp 357–374
Hutchinson JR, Garcia M (2002) *Tyrannosaurus* was not a fast runner. Nature 415:1018–1021
Jablonski D (1989) The biology of mass extinction: a palaeontological view. Philos Trans R Soc Lond B 325:357–368
Jenson JA (1985) Three new sauropod dinosaurs from the Upper Jurassic of Colorado. Great Basin Nat 45:697–709
Jepson G L (1964) Riddles of the terrible lizards. Am Sci 52:227–246
Ji Q, Currie PJ, Ji S, Norell MA (1998) Two feathered dinosaurs from northeastern China. Nature 393:753–761
Ji Q, Norell MA, Gao K-Q, Ji S-A, Ren D (2001) The distribution of integumentary structures in a feathered dinosaur. Nature 410:1084–1088
Kemp TS (1982) Mammal-like reptiles and the origin of mammals. Academic Press, London, xiv + 363 pp
King G (1996) Reptiles and herbivory. Chapman and Hall, London, vii + 160 pp
Kurtén B (1968) The age of dinosaurs (World University Library). Weidenfeld and Nicholson, London, 255 pp
Kurtén B (1969) Continental drift and evolution. Sci Am 2265(3):54–64
Lambert D and the Diagram Group (1992) Dinosaur data book. The definitive fully illustrated encyclopedia of dinosaurs and other prehistoric reptiles. Facts on File, New York, 320 pp
Langston W Jr (1981) Pterosaurs. Sci Am 244(2):92–102
Lawson DA (1975) Pterosaur from the latest Cretaceous of West Texas: discovery of the largest flying creature. Science 187:947–948
Lockley MG (1997) The paleoecological and paleoenvironmental utility of dinosaur tracks. In: Farlow JO, Brett-Surman MK (eds) The complete dinosaur. Indiana Univ Press, Bloomington, pp 554–578
MacLean N (1996) K-T redux. Paleobiology 22:311–317
Maryańska T (1997) Segnosaurs (Therizinosaurs). In: Farlow JO, Brett-Surman MK (eds) The complete dinosaur. Indiana Univ Press, Bloomington, pp 234–241
Massare JA (1988) Swimming capabilities of Mesozoic marine reptiles: implications for method of predation. Paleobiology 14:187–205
Massare JA (1997) Faunas, behavior and evolution. Introduction. In: Callaway JM, Nicholls EL (eds) Ancient marine reptiles. Academic Press, San Diego, pp 401–421
McGowan C (1972) The systematics of Cretaceous ichthyosaurs with particular reference to the material from North America. Contrib Geol 11:9–29
McGowan C (1991) Dinosaurs, Spitfires and sea dragons. Harvard Univ Press, Cambridge, x + 365 pp
McIntosh JS, Brett-Surman MK, Farlow JO (1997) Sauropods. In: Farlow JO, Brett-Surman MK (eds) The complete dinosaur. Indiana Univ Press, Bloomington, pp 264–290
Milne A (1991) The fate of the dinosaurs. New perspectives in evolution and extinction. Prism Press, Bridport Dorset, ix + 301 pp
Milner A (2002) Dyno-birds. From dinosaurs to birds. The Natural History Museum, London, 64 pp

Molnar RE, Farlow JO (1990) Carnosaur paleobiology. In: Weishampel DB, Dobson P, Osmólska H (eds) The Dinosauria. Univ California Press, Berkeley, pp 210–224
Moody R (1977) A natural history of dinosaurs (Ilustr Middleton J, Parry L, Morris T). Hamlyn, London, 124 pp
Morales M (1997) Major groups of non-dinosaurian vertebrates of the Mesozoic Era. In: Farlow JO, Brett-Surman MK (eds) The complete dinosaur. Indiana Univ Press, Bloomington, pp 607–626
Naish O, Martill DM (2003) Pterosaurs – a successful invasion of prehistoric skies. Biologist 50(5):213–216
Neill WT (1971) The last of the ruling reptiles. Alligators, crocodiles, and their kin. Columbia Univ Press, New York, xvii + 486 pp
Norman D (1985) The illustrated encyclopedia of dinosaurs (Illstr Sibbick J) Salamander Books, London, 208 pp (incorporated in Norman and Wellnhofer 2000)
Norman D (1991) Dinosaur! Boxtree, London, 288 pp
Norman D, Wellnhofer P (2000) The illustrated encyclopedia of dinosaurs. An original and compelling insight into life in the dinosaur kingdom (Illustr Sibbick J). Salamander Books, London, 400 pp
Norrell MA, Clark JM, Chiappe LM, Dashzveg D (1995) A nesting dinosaur. Nature 378:774–776
Ostrom JH (1964) A reconstruction of the paleoecology of hadrosaurian dinosaurs. Am J Sci 262:975–997
Ostrom JH (1969) Terrestrial vertebrates as indicators of Mesozoic climates. Proc North Am Paleontol Convention, pp 347–376
Ostrom JH (1972) Were some dinosaurs gregarious? Palaeogr Palaeoclim Palaeoecol 11:283–301
Ostrom JH (1974a) *Archaeopteryx* and the origin of flight. Q Rev Biol 49:27–47
Ostrom JH (1974b) Reply to 'dinosaurs as reptiles'. Evolution 28:491–493
Padian K (1984) A functional analysis of flying and walking in pterosaurs. Paleobiology 9:218–239
Padian K (ed) (1986) The beginning of the age of dinosaurs. Faunal change across the Triassic-Jurassic boundary. Cambridge Univ Press, Cambridge, xii + 378 pp
Padian K (1997) Bipedality. In: Currie PJ, Padian K (eds) Encyclopedia of dinosaurs. Academic Press, San Diego, pp 68–70
Padian K, Chiappe LM (1997) Bird origins. In: Currie PJ, Padian K (eds) Encyclopedia of dinosaurs. Academic Press, San Diego, pp 71–79
Padian K, Rayner JMV (1993) The wings of pterosaurs. Am J Sci 293A:91–166
Palmer D (ed) (1999) The illustrated encyclopedia of dinosaurs and prehistoric animals (Reptiles pp 58–169 Illustr Kirk S). Marshall, London, 312 pp
Parker LR, Rowley RL Jr (1989) Dinosaur footsteps from a coal mine in east-central Utah. In: Gillette DD, Lockley MG (eds) Dinosaur tracks and traces. Cambridge University Press, New York, pp 361–366
Parrish JM (1997) Evolution of the archosaurs. In: Farlow JO, Brett-Surman MK (eds) The complete dinosaur. Indiana Univ Press, Bloomington, pp 191–203
Paul GS (1997) Reproductive behavior and rates. In: Currie PJ, Padian K (eds) Encyclopedia of dinosaurs. Academic Press, San Diego, pp 630–637
Pritchard PCH (1979) Encyclopedia of turtles. TFH Publications, Neptune New Jersey, Reigate Surrey, 895 pp
Raup DM (1989) The case for extraterrestrial causes of extinction. Philos Trans R Soc Lond B 325:421–435
Raup OM (1992) Extinction, bad genes or bad luck? WW Norton, London, 210 pp
Reid REH (1984) The histology of dinosaurian bone and its possible bearing on dinosaurian physiology. In: Ferguson MWJ (ed) The structure, development and evolution of reptiles. Symp Zool Soc Lond 52:629–663
Reid REH (1997) Dinosaurian physiology: the case for 'intermediate' dinosaurs. In: Farlow JO, Brett-Surman MK (eds) The complete dinosaur. Indiana Univ Press, Bloomington, pp 449–473
Robinson JA (1975) The locomotion of plesiosaurs. N Jahrb Geol Paläontol Abh 149:286–232
Rodbard S (1949) On the dorsal sail of *Dimetrodon*. Copeia 1949:224

Romer AS (1948) Relative growth in pelycosaurian reptiles. Spec Publ R Soc S Afr, R Broom Commemoration vol, pp 45–55

Romer AS (1966) Vertebrate paleontology, 3rd edn. Univ Chicago Press, Chicago, iv + 469 pp

Ross C A, Garnett S (eds) (1989) Crocodiles and alligators. Welden Owen, McMahons Point NSW, Merehurst Press, London, 240 pp

Rothschild B M (1997) Dinosauran paleopathology. In: Farlow JO, Brett-Surman MK (eds) The complete dinosaur. Indiana Univ Press, Bloomington, pp 426–448

Ruben J, Leitch A, Hillenius W, Geist N, Jones T (1997) New insights into the metabolic physiology of dinosaurs. In: Farlow JO, Brett-Surman MK (eds) The complete dinosaur. Indiana Univ Press, Bloomington, pp 504–518

Russell D (1971) The disappearance of the dinosaurs. Can Geog J 83:204–215

Russell DA, Dodson P (1997) The extinction of the dinosaurs: a dialogue between a catastrophist and a gradualist. In: Farlow JO, Brett-Surman MK (eds) The complete dinosaur. Indiana Univ Press, Bloomington, pp 662–672

Russell LS (1965) Body temperature of dinosaurs and its relationships to their extinction. J Palaeontol 39:497–501

Ryan MJ, Vickaryous MK (1997) Diet. In: Currie PJ, Padian K (eds) Encyclopedia of dinosaurs. Academic Press, San Diego, pp 169–174

Sampson S (1997) Dinosaur combat and courtship. In: Farlow JO, Brett-Surman MK (eds) The complete dinosaur. Indiana Univ Press, Bloomington, pp 383–393

Schatz A (1957) Some biochemical and physiological considerations regarding the extinction of the dinosaurs. Proc Penn Acad Sci 32:26–36

Schmidt-Nielsen K (1984) Scaling: why is animal size so important? Cambridge Univ Press, Cambridge, xi + 241 pp

Sereno PC, Forster CA, Rogers RR, Monetta A (1993) A primitive dinosaur skeleton from Argentina and the early evolution of Dinosauria. Nature 361:64–66

Seymour RC (1976) Dinosaurs, endothermy and blood pressure. Nature 262:207–208

Spicer RA, Parrish JT (1990) Late Cretaceous–Early Tertiary paleoclimates of northern high latitudes: a quantitative view. J Geol Soc Lond 147:329–341

Špinar ZV (Benton MJ ed) (1995) Life before man (illustr Burian Z) (Transl Kadečka S, Schierlová M) (new edn). Thames and Hudson, London, 256 pp

Spotila JR, Lommen PW, Bakhen GS, Gates DM (1973) A mathematical model for body temperatures of large reptiles: implications for dinosaur ecology. Am Nat 107:391–404

Stieler C (1922) Neuer Rekonstruktionsversuck eines liaissischn Flugsauriers. Naturwiss Wochenschr NF 21(20):273–280

Swain T (1976) Angiosperm-reptile coevolution. In: Bellairs A d'A, Cox CB (eds) Morphology and biology of reptiles. Academic Press, London, pp 107–122

Taylor MA (1986) Lifestyle of plesiosaurs. Nature 319:179

Taylor MA (1987a) A reinterpretation of ichthyosaur swimming and buoyancy. Palaeontology 30:531–535

Taylor MA (1987b) How tetrapods feed in water: a functional analysis by paradigm. Zool J Linn Soc 91:171–195

Taylor MA (1992) Functional anatomy of the large aquatic predator *Rhomaleosaurus zetlandicus* (Plesiosauria, Reptilia) from the Toarcian (Lower Jurassic) of Yorkshire, England. Philos Trans R Soc Lond (B) 335:247–280

Thomas RDK, Olson EC (eds) (1980) A cold look at the warm-blooded dinosaurs. AAAS selected symposium 28. Westview Press, Boulderco, 514 pp

Thulborn RA (1973) Thermoregulation in dinosaurs. Nature 245:51–52

Thulborn R A (1975) Dinosaur polyphyly and the classification of archosaurs and birds. Aust J Zool 23:249–270

Thulborn T (1978) Aestivation among ornithopod dinosaurs of the African Trias. Lethaia 11:185–198

Thulborn T (1990) Dinosaur tracks. Chapman and Hall, London, xvii + 409 pp

Tiffney B H (1997) Land plants as food and habitat in the age of the dinosaurs. In: Farlow JO, Brett-Surman MK (eds) The complete dinosaur. Indiana Univ Press, Bloomington, pp 352–370
Toon O B, Zahnle K, Morrison P, Turco R, Covey C (1997) Environmental perturbations caused by the impacts of asteroids and comets. Rev Geophys 35:41–78
Tracy CR, Turner JS, Huey RB (1986) A biophysical analysis of possible thermoregulatory adaptations in sailed pelycosaurs. In: Hotton N III, Maclean PD, Roth JJ, Roth EC et al. (eds) The ecology and biology of mammal-like reptiles. Smithsonian Institute Press, Washington, DC, pp 195–206
Unwin DM (1988) Extinction and survival in birds. In: Larwood GP (ed) Extinction and survival in the fossil record. Systematics Association special volume no 24. Clarendon Press, Oxford, pp 295–318
Unwin DM, Bakhurina NN (1994) *Sordes pilosus* and the nature of the pterosaur flight apparatus. Nature 371:62–64
Urey HC (1973) Cometary collisions and geological periods. Nature 242:32–33
VanHeerden J (1997) Prosauropods. In: Farlow JO, Brett-Surman MK (eds) The complete dinosaur. Indiana Univ Press, Bloomington, pp 242–263
Vickaryous MK, Ryan MJ (1997) Ornamentation. In: Currie PJ, Padian K (eds) Encyclopedia of dinosaurs. Academic Press, San Diego, pp 488–493
Wall WP (1983) The correlation between high limb-bone density and aquatic habitats in Recent mammals. J Palaeontol 57:197–207
Weishampel DB, Dobson P, Osmólska H (eds) (1990) The Dinosauria. Univ California Press, Berkeley, xxii + 733 pp
Wellnhofer P (1975) Die Rhamphorhynchoidea (Pterosauria) der Oberjura–Plattenkalke Süddeutschlands. Palaeontographica A 148:1–33
Wellnhofer P (1991) The illustrated encyclopedia of pterosaurs (Illustr Sibbick J). Salamander Books, London, Crescent Books, New York, 192 pp (incorporated in Norman and Wellnhofer 2000)
Wild R (1978) Die Triasfauna des Tessiner Kalkalpen xxiii. *Tanystropheus longobardicus* (Bassani) (Neue Ergebnisse). Schw Palaöntol Abh 95:1–162
Wild R (1984) Flugsaurier aus der Obertrias von Italien. Naturwissenschaften 71: 1–11
Wilford JN (1986) The riddle of the dinosaur. Knopf, New York, 304 pp
Young JZ (1981) The life of vertebrates, 3rd edn. Clarendon Press, Oxford, xv + 645 pp

Subject Index

Page numbers in *italics* refer to illustrations

Printing: Krips bv, Meppel
Binding: Litges & Dopf, Heppenheim